my revision notes

Edexcel International GCSE (9–1)
CHEMISTRY

Neil Dixon

HODDER
EDUCATION
AN HACHETTE UK COMPANY

The publisher would like to thank the following for permission to reproduce copyright material:

Photo credits: p. 2 © geejays/123RF, page 110 (four photos) © antonio scarpi – Fotolia.com

Although every effort has been made to ensure that website addresses are correct at time of going to press, Hodder Education cannot be held responsible for the content of any website mentioned. It is sometimes possible to find a relocated web page by typing in the address of the home page for a website in the URL window of your browser.

Orders: please contact Bookpoint Ltd, 130 Milton Park, Abingdon, Oxon OX14 4SB. Telephone: (44) 01235 827720. Fax: (44) 01235 400401. Lines are open 9.00–17.00, Monday to Saturday, with a 24-hour message answering service. Visit our website at www.hoddereducation.co.uk

Cover photo © sbarabu – stock.adobe.com

Illustrations by Aptara, Inc.

Typeset in BemboStd 11/13 pts by Aptara Inc.

Printed in Spain

ISBN 9781510446748

Get the most from this book

Everyone has to decide his or her own revision strategy, but it is essential to review your work, learn it and test your understanding. These Revision Notes will help you to do that in a planned way, topic by topic. Use this book as the cornerstone of your revision and don't hesitate to write in it – personalise your notes and check your progress by ticking off each section as you revise.

Tick to track your progress

Use the revision planner on pages iv and v to plan your revision, topic by topic. Tick each box when you have:

- revised and understood a topic
- tested yourself
- practised the exam questions and gone online to check your answers and complete the quick quizzes.

You can also keep track of your revision by ticking off each topic heading in the book. You may find it helpful to add your own notes as you work through each topic.

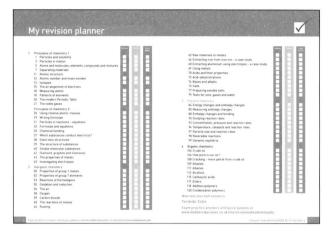

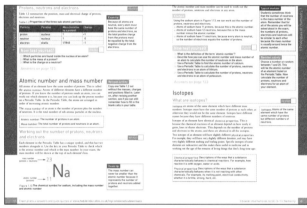

Features to help you succeed

Exam tips

Expert tips are given throughout the book to help you polish your exam technique in order to maximise your chances in the exam.

Typical mistakes

The author identifies the typical mistakes candidates make and explain how you can avoid them.

Now test yourself

These short, knowledge-based questions provide the first step in testing your learning. Answers are at the back of the book.

Definitions and key words

Clear, concise definitions of essential key terms are provided where they first appear.

Key words from the specification are highlighted in bold throughout the book.

Revision activities

These activities will help you to understand each topic in an interactive way.

Exam practice

Practice exam questions are provided for each topic. Use them to consolidate your revision and practise your exam skills.

Summaries

The summaries provide a quick-check bullet list for each topic.

Online

Go online to check your answers to the exam questions and try out the extra quick quizzes at **www.hoddereducation.co.uk/myrevisionnotes downloads**

My revision planner

Exam practice answers and quick quizzes at **www.hoddereducation.co.uk/myrevisionnotesdownloads**

Exam practice answers and quick quizzes at

www.hoddereducation.co.uk/myrevisionnotesdownloads

Countdown to my exams

6–8 weeks to go

- Start by looking at the specification — make sure you know exactly what material you need to revise and the style of the examination. Use the revision planner on pages iv and v to familiarise yourself with the topics.
- Organise your notes, making sure you have covered everything on the specification. The revision planner will help you to group your notes into topics.
- Work out a realistic revision plan that will allow you time for relaxation. Set aside days and times for all the subjects that you need to study, and stick to your timetable.
- Set yourself sensible targets. Break your revision down into focused sessions of around 40 minutes, divided by breaks. These Revision Notes organise the basic facts into short, memorable sections to make revising easier.

REVISED ☐

2–6 weeks to go

- Read through the relevant sections of this book and refer to the exam tips, summaries, typical mistakes and key terms. Tick off the topics as you feel confident about them. Highlight those topics you find difficult and look at them again in detail.
- Test your understanding of each topic by working through the 'Now test yourself' questions in the book. Look up the answers at the back of the book.
- Make a note of any problem areas as you revise, and ask your teacher to go over these in class.
- Look at past papers. They are one of the best ways to revise and practise your exam skills. Write or prepare planned answers to the exam practice questions provided in this book. Check your answers online and try out the extra quick quizzes at **www.hoddereducation.co.uk/ myrevisionnotesdownloads**
- Use the revision activities to try out different revision methods. For example, you can make notes using mind maps, spider diagrams or flash cards.
- Track your progress using the revision planner and give yourself a reward when you have achieved your target.

REVISED ☐

One week to go

- Try to fit in at least one more timed practice of an entire past paper and seek feedback from your teacher, comparing your work closely with the mark scheme.
- Check the revision planner to make sure you haven't missed out any topics. Brush up on any areas of difficulty by talking them over with a friend or getting help from your teacher.
- Attend any revision classes put on by your teacher. Remember, he or she is an expert at preparing people for examinations.

REVISED ☐

The day before the examination

- Flick through these Revision Notes for useful reminders, for example the exam tips, exam summaries, typical mistakes and key terms.
- Check the time and place of your examination.
- Make sure you have everything you need — extra pens and pencils, tissues, a watch, bottled water, sweets.
- Allow some time to relax and have an early night to ensure you are fresh and alert for the examinations.

REVISED ☐

My exams

Edexcel International GCSE Chemistry Paper 1

Date:..

Time:..

Location:..

Edexcel International GCSE Chemistry Paper 2

Date:..

Time:..

Location:..

1 Principles of chemistry 1

Particles and solubility

All substances are made of particles. Sometimes you can see particles with your eye, like grains of sand or individual crystals of salt. However, when chemists use the term 'particles' they are usually talking about **atoms**, **ions** or **molecules**. These particles are far too small to be seen without the most powerful electron microscopes.

> **Atom:** The smallest particle of an element, with a central nucleus surrounded by orbiting electrons.
>
> **Ion:** A charged particle, which is usually formed from an atom that has gained or lost electrons.
>
> **Molecule:** A cluster of non-metal atoms which are joined by strong covalent bonds.

Particles in solutions

REVISED

A **solution** is a mixture of a liquid with another substance, which is called the **solute**. The solute is usually a solid, but it can also be a gas. When a solute dissolves to make a solution, it separates into individual particles that are too small to see and too small to settle out at the bottom of the liquid. The liquid which makes up most of the volume of the solution is the **solvent**. Water is the most common solvent, but not all substances are soluble in water, though they may be **soluble** in other solvents. When a substance will not dissolve in a particular solvent, it is said to be **insoluble**.

A **saturated solution** is formed when enough solute has been added to a solvent so that no more solute will dissolve. The amount of solute that dissolves in a specific volume of a solvent is called the **solubility**. Solubility is usually measured in 'g of solute per 100 g of water'.

> **Solution**: A mixture formed from a solvent (liquid) and a solute, where the solute is present as individual particles.
>
> **Solute**: The substance that has been dissolved in a solvent to make a solution.
>
> **Solvent**: The liquid that makes up most of the volume of a solution.
>
> **Soluble**: A substance is soluble in a particular solvent if it dissolves in it to form a solution.
>
> **Insoluble**: A substance is insoluble in a solvent if it does not dissolve in it.
>
> **Saturated solution**: A solution which cannot dissolve any more solvent.
>
> **Solubility**: The mass of a solute that will dissolve in a specified mass of solvent.

Revision activity

Write each key term from this section onto a separate card and write the relevant definition on the other side. With the definitions facing down, pick a card at random and test yourself on its definition. Then turn over the card to see if you are right.

Required practical

Investigate the solubility of a solid in water at a specific temperature

Measure the maximum mass of a solid that will dissolve in 100 g of water at a specific temperature.

1. 100 cm^3 of water is measured and poured into a beaker. 100 cm^3 water has a mass of 100 g.
2. The temperature of the water is recorded. A thermostatically controlled water bath can be used to investigate solubility at temperatures above room temperature.
3. 100 g of the solute is weighed into a clean, dry beaker.

4 One spatula of the solute is added to the water, which is stirred using a glass rod.

5 When the solute has dissolved, more of the solute is added, one spatula at a time, with stirring. No more solute is added when some remains undissolved after stirring.

6 The mass of solute that remains unused is recorded.

7 The mass of unused solute is subtracted from the starting mass (100 g) to calculate the mass of solute that dissolved in 100 g of water at the specific solvent temperature. This is the solubility.

8 If different students investigate the solubility of one solute at different temperatures, the results can be combined and then a solubility curve can be plotted, like the one in Figure 1.1.

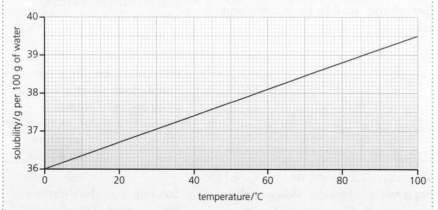

Figure 1.1 A graph of the solubility of sodium chloride at different temperatures

You can interpret this graph in several ways.

● General conclusion: as the temperature increases, the solubility of sodium chloride increases.

● You can see that the solubility of sodium chloride at 80 °C is 38.8 g per 100 g of water, because you can find 80 °C on the x-axis and then follow the vertical gridline up until you reach the red line, then follow a horizontal gridline until you reach 38.8 g on the y-axis.

● You can calculate that 388 g of sodium chloride would dissolve in one litre of water (1000 cm³ or 1000 g) at 80 °C because you know that 38.8 g will dissolve in 100 cm³ of water, and therefore you can scale this up by a factor of ten.

Diluting coloured solutions

REVISED

Many solutes are white, like sodium chloride (common salt) so they make colourless solutions. But some solutes are coloured, and they make coloured solutions. For example, copper sulfate normally exists as blue crystals, and when they are dissolved in water, a clear blue solution is made.

When water is added to a coloured solution, the colour becomes paler. This is because the particles which are causing the colour are becoming more spread out.

Figure 1.2 Copper sulfate solution

Typical mistake

When describing solutions, do not confuse the term **clear** with **colourless**. **Clear** means that you can see through it, but it might still have a colour. If you mean that it looks like water, use the term **colourless**.

TESTED ☐

Now test yourself

1 What are the differences between atoms, ions and molecules?
2 What word is used for the liquid in a solution?
3 What are the most common units for solubility?
4 What is meant by the term 'saturated solution'?

Answers on page 123

Particles in motion

States of matter

REVISED ☐

The kinetic theory of matter describes the arrangement and movement of the tiny particles that make up solids, liquids and gases. The kinetic theory is a useful model because it explains the properties of solids, liquids and gases.

Table 1.1 A comparison of solids, liquids and gases

	Solid	Liquid	Gas
Diagram			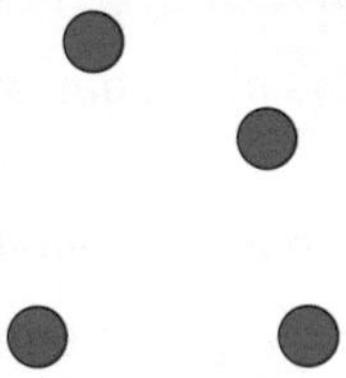
Arrangement of particles and relevant properties	Regularly arranged, so solids often exist as crystals. Touching neighbouring particles, so solids are usually denser than liquids and gases.	Randomly arranged. Mostly touching the neighbouring particles, so liquids cannot usually be compressed.	Randomly arranged. Widely spaced, with nothing in between, so gases usually have a low density and can be easily compressed.
Movement of particles and relevant properties	Fixed in place, and move only by vibrating, so solids have a fixed shape.	Can move over each other, so liquids can be poured and take the shape of their container.	Move very quickly, so gases take up the space of their container, and diffuse easily.
Energy of particles	Low	Medium	High

Typical mistake

Students sometimes draw the particles in a liquid with small spaces between all of the particles. This cannot be correct, otherwise liquids would be compressible. Remember instead to draw them so they are mostly touching but randomly arranged.

Revision activity

Copy the key facts and the diagrams from Table 1.1, onto separate sticky notes. Shuffle them and then sort the characteristics back into the three categories: solid, liquid, gas.

Changes of state

REVISED ☐

Most pure substances can exist as a solid, liquid or gas, depending on the temperature. The changes of state that occur between them are summarised in Figure 1.3.

In Figure 1.3, the red arrows represent interconversions which are caused by heating and the blue arrows represent changes of state which are caused by cooling.

Heating a solid gives the particles more energy, so they vibrate faster and eventually break free of the bonds holding them in place. This means that the solid is **melting**. Further heating gives the particles even more energy, and so particles overcome the attractive forces in the liquid, and this is called evaporation or boiling. **Evaporating** is when a liquid is slowly turning into a gas, when temperatures are below the **boiling point**. When a liquid is heated to its boiling point, it starts turning into a gas rapidly, causing bubbles to form. This is called **boiling**.

Cooling a gas down causes the particles to lose energy. They no longer have enough energy to remain widely spaced, so they come closer together and form a liquid. This is called **condensing**. Further cooling causes the particles to lose even more energy and arrange themselves regularly, which is called **freezing**.

Gases can also be made to condense by increasing the pressure, because this forces the particles closer together until they are mostly touching, which means the gas has condensed.

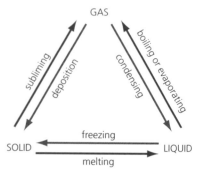

Figure 1.3 Changes of state

Melting: When a solid turns into a liquid.

Evaporating: When a liquid slowly turns into a gas at temperatures lower than the boiling point.

Boiling point: The temperature at which a pure liquid boils.

Boiling: When a liquid quickly turns into a gas at temperatures higher than the boiling point.

Condensing: When a gas turns into a liquid.

Freezing: When a liquid turns into a solid.

Diffusion

REVISED

Diffusion occurs when particles from one substance spread through another substance from an area of high concentration to an area of lower concentration. Examples of diffusion include:
- Being able to smell something from a few metres away.
- Adding a little coloured solution to a beaker of water and eventually the solution becoming evenly coloured.

Diffusion occurs quickly through the air because the particles in gases are widely spaced and moving very quickly. They bump into the particles that are causing the smell and knock them around randomly. The smell therefore spreads out quickly.

Diffusion occurs much more slowly through water because the particles in a liquid are much closer together and move more slowly.

Diffusion does not usually occur in solids.

Now test yourself

5 What change of state occurs when water is cooled from 150 °C to 70 °C?
6 Copy and complete the following table on the three states of matter.

	Solid	Liquid	Gas
Arrangement of particles	Regularly arranged, touching neighbours		
Movement of particles		Can move over each other	Moving fast

7 What process is occurring when a puddle dries up on a warm day?
8 What process explains why snow disappears on a sunny winter day without melting into liquid water?

Answers on page 123

Atoms and molecules; elements, compounds and mixtures

Atoms and molecules

An **atom** is the smallest particle of an **element**. Scientists used to think that atoms could not be split, but we now know that atoms are made from even smaller particles called protons, neutrons and electrons (see page 11).

Sometimes it is still helpful to think of (and draw) atoms as small spheres. We sometimes give atoms of different elements a specific colour to help us better understand diagrams of molecules.

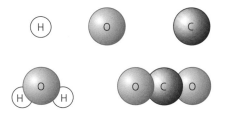

Figure 1.4 Top row: atoms of hydrogen, oxygen and carbon. Bottom row: a molecule of water and a molecule of carbon dioxide

A **molecule** is a cluster of non-metal atoms which are chemically bonded to each other via **covalent** chemical bonds. Some non-metal elements form molecules, like oxygen (O_2) and nitrogen (N_2). Other non-metal elements do not form molecules, like silicon (Si), so there are no numbers in the formula for silicon. Most compounds formed from two or more non-metals are made of molecules. Examples include water (H_2O), methane (CH_4) and ammonia (NH_3). The molecular formula tells you how many of each type of atom are present in a molecule.

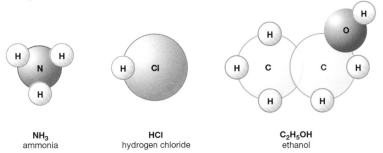

NH_3
ammonia

HCl
hydrogen chloride

C_2H_5OH
ethanol

Figure 1.5 Diagrams and molecular formulae can be used to show how many of each type of atom are present in a molecule

Atom: The smallest particle of an element.

Element: A pure chemical substance that is listed on the Periodic Table (see page 130). Elements cannot be broken down into simpler substances by chemical reactions.

Molecule: A cluster of two or more non-metal atoms which are joined by covalent bonds.

Covalent bond: A type of chemical bond which occurs between two non-metal atoms which share electrons (see page 36).

Typical mistake

Students are sometimes careless with their use of capital and lowercase letters when writing chemical symbols. The symbol for an element always starts with a capital letter. If there is a second letter, it is always lowercase. For example, C is carbon, O is oxygen. So CO is carbon monoxide. Co is a different element – cobalt. cO has no meaning, and nor does co.

Elements, compounds and mixtures

An element is a pure substance which is made of one type of atom. We use a unique chemical symbol for each element. The elements are all listed on the Periodic Table (see page 130) and you can use this to find the symbol for any element.

A **compound** is a pure chemical substance made from two or more elements which are chemically bonded to each other in a fixed ratio. This means that a compound has a specific chemical formula, but a mixture does not. For example, nitrogen will bond to oxygen to make nitrogen dioxide (NO_2) but it can also form a different compound with oxygen called nitrogen monoxide (NO). A pure compound always has a fixed melting point and boiling point.

A **mixture** is an impure substance made from more than one element which are not chemically bonded, or from more than one compound. For example, air is a mixture of the elements nitrogen, oxygen and argon, and the compound carbon dioxide, plus other compounds in tiny amounts. Mixtures are usually easier to separate (see page 7) into their constituent compounds and elements than compounds are, because no chemical reactions are needed during these separation processes.

The difference between pure and impure substances

A **pure** substance is an element or a compound which contains only one chemical substance. A pure **element** is made of only one type of atom, although these atoms might be bonded into molecules, such as chlorine, which is made of Cl_2 molecules.

A pure **compound** does contain more than one type of atom, but these are bonded together. For example, pure water contains only molecules which are each made of two atoms of hydrogen chemically bonded to one atom of oxygen.

If a substance contains small amounts of another chemical, it is described as impure and the unwanted substance in the mixture is called the **impurity**. For example, pure gold is more valuable than impure gold, which may contain copper and tin impurities.

A pure substance will have a fixed melting point and boiling point. These values can be found in data books or from reliable online sources. To check if a substance is pure, you can heat it up and record the temperature that it melts or boils and then compare it with the correct value.

If a substance is impure, then it will not melt or boil at the expected temperature. An impure substance is also likely to melt or boil over a range of temperatures.

> **Compound:** A pure substance which is made from more than one element that are chemically bonded together in a fixed ratio.
>
> **Mixture:** An impure substance that is made from more than one element or compound, which are not chemically bonded together.

> **Typical mistake**
>
> Metal elements do not form molecules because they do not form covalent bonds, so when representing a metal in a symbol equation, do not include a number. For example, sodium is Na, not Na_2.

> **Pure:** A substance that contains only one compound or element, and does not have any impurities.
>
> **Impurity:** An unwanted substance which is present in a mixture.

Now test yourself

9 Use a Periodic Table to find the symbol for potassium.
10 The formula for butane is C_4H_{10}. Is it an element or a compound?
11 Look at the formula $C_2H_4O_2$. How many of each type of atom are present and what are they?
12 Why does air have no chemical formula?

Answers on page 123

Separating materials

Filtration

The apparatus for filtration is shown in Figure 1.6. It is used to separate an **insoluble** solid from a liquid. Filtration works because the particles of the solid are much larger than the molecules and ions in a liquid or solution. This means that the molecules and ions can pass through very small holes in the filter paper to form the **filtrate**, but the larger particles of insoluble solid cannot, so they remain as a **residue** in the filter paper.

Insoluble: A substance that will not dissolve in a specific solvent.

Filtrate: The liquid which is formed after filtration. The filtrate does not contain any insoluble solid.

Residue: The solid which is separated during filtration. It remains in or on the filter paper.

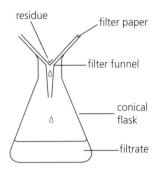

Figure 1.6 The apparatus used for filtration in the lab

Crystallisation

To obtain a solid solute from a solution, **evaporation** and **crystallisation** are used. These processes remove the unwanted solvent. Here are the steps in evaporation and crystallisation:

- Some solution is poured into an evaporating basin until it is half full.
- The evaporating basin is heated gently over a beaker of boiling water which is on a gauze resting on a tripod over a Bunsen burner. The steam will heat the bottom of the evaporating basin to 100 °C.
- When the volume of the solution in the basin has halved, the gas is turned off and the apparatus is allowed to cool.
- The evaporating basin is moved to a warm place and left for a few days for crystals to form in the remaining liquid.
- The crystals are removed with tweezers and dried between some paper towels.

It is possible to evaporate the solvent from a solution more quickly by heating the evaporating basin directly with a Bunsen flame, but the method described above produces larger crystals and is usually safer.

Evaporation: When a liquid turns into a gas.

Crystallisation: When crystals form from a solution, typically after the volume of the solution has been reduced and allowed to cool.

evaporating basin
copper sulfate solution
boiling water
beaker (water bath)

HEAT

heat-proof mat

Figure 1.7 The apparatus used to produce crystals of copper sulfate from copper sulfate solution

Simple distillation

In some circumstances, a solution must be purified to obtain the solvent, when the solutes are the unwanted part. An example of this is the way that purified water is obtained from tap water, which contains dissolved solids and gases. Simple **distillation** involves heating the solution until it boils and then cooling the gas produced until it condenses. The impurities are left behind in the original solution.

> **Distillation**: When a solvent is boiled and then condensed to obtain the pure solvent.

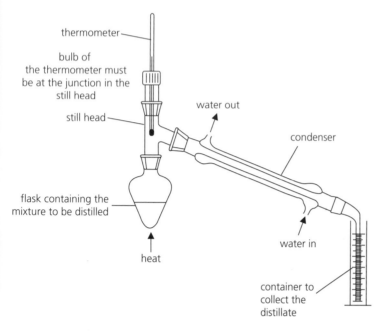

Typical mistake

Some students draw a diagram of the apparatus for distillation with the water entering the top of the condenser. It is essential that the cooling water for the condenser enters the lower part of the condenser and leaves via the top pipe, as indicated in the diagram.

Figure 1.8 The apparatus used to distil a solution and collect the solvent

The apparatus and method shown in Figure 1.8 can also be used to separate two liquids which have different boiling points. For example, a mixture of ethanol and water can be distilled in this way. By warming the mixture until the temperature on the thermometer is 79 °C (just above ethanol's boiling point), the ethanol boils off and is condensed, while most of the water remains in the flask. It is worth noting that some of the water will have evaporated, even at 79 °C, so the ethanol distillate will not be pure.

Fractional distillation

Some liquids are complex mixtures of more than two compounds, which may all have different boiling points. Where this is the case, **fractional distillation** can be used to separate all of the different liquids in the mixture. Fractional distillation of a mixture of ethanol and water is more likely to produce a pure distillate of ethanol.

During fractional distillation, the flask is heated gently until a liquid starts to condense in the condenser. This liquid is the first **fraction**. The temperature indicated on the thermometer should be recorded because this is the boiling point of the first fraction. The heating used on the flask is then maintained so the temperature at the top of the column remains constant until no more of the first fraction is produced.

> **Fractional distillation**: A separation technique that separates liquids from a mixture on the basis of their different boiling points.
>
> **Fraction**: A liquid which has been separated from a mixture using fractional distillation.

Then the container used to collect the first fraction is replaced. Heating is then gradually increased until the second fraction starts to distil off. Again, the new (higher) temperature is maintained until the second fraction is completely collected. The third, fourth and any further fractions can also be collected in this way, with new containers each time and successively higher temperatures.

Fractional distillation can be used to separate the mixture of liquids in crude oil in the laboratory. However, when fractional distillation of crude oil is done on an industrial scale, the apparatus and method are quite different (see page 104–5).

Paper chromatography

REVISED

Chromatography can be used to separate mixtures of substances that are all soluble in a specific solvent, but which have different solubilities. For example, it can be used to separate coloured pigments in a water-soluble pen ink, or in food colourings. It can also be used to test the purity of medicines.

In paper chromatography, the mixture is separated out when the solvent (typically water) moves up a piece of paper. The separation occurs because some pigments are more strongly attracted to the water and only weakly attracted to the paper. These pigments will move fastest up the paper. Other pigments are strongly attracted to the paper and only weakly attracted to the water molecules. These pigments will move slowly up the paper, and at the end of the experiment will be lower down the chromatogram.

In chromatography, the chemical which moves, i.e. the solvent (usually water) is called the **mobile phase**. The chemical which stays still, i.e. the paper, is called the **stationary phase**.

While it is most common to use chromatography to separate coloured pigments in a mixture, it can also be used for colourless substances. After the experiment, the chromatogram is sprayed with a chemical called a 'locating agent' which reacts with the substances that are now separated and makes them visible.

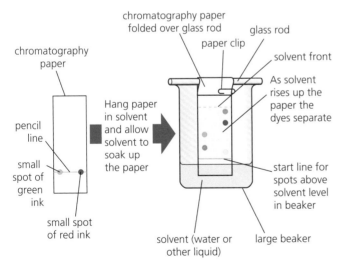

Figure 1.10 The apparatus used for paper chromatography to separate the pigments in a green ink and a red ink

From the results at the end of the experiment in Figure 1.10, we can see that the green ink contained two pigments (blue and green). We can also see that the red ink contained yellow, red and orange pigments. No pigment was present in both of the original inks. The orange pigment

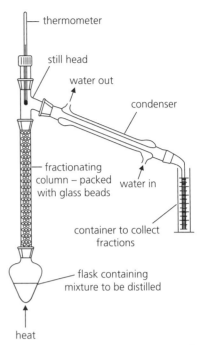

Figure 1.9 The apparatus used for fractional distillation

Chromatography: A method used to separate a mixture of soluble substances.

Mobile phase: The solvent which carries the substances to be separated.

Stationary phase: The substance which remains still during chromatography.

Exam tip

You must state in the exam that the start line is drawn in pencil. This is important because if the line is drawn in pen, the line of ink will move up the paper as the solvent rises. You also need to ensure that the level of the solvent in the beaker is below the start line at the beginning of the experiment. If the spots of ink are submerged in the solvent, they will wash off.

had the greatest attraction for the solvent, and the yellow ink had the greatest affinity for the paper.

Calculating R_f values

In order to compare one chromatogram with another in a useful way, you can calculate the R_f value for each of the spots. To do this, you must mark the level reached by the solvent as soon as you remove the chromatogram from the solvent beaker. Speed is important because the solvent will dry up off the paper quite quickly and you will not be able to see where the solvent reached. R_f values are calculated using the following formula.

$$R_f = \frac{\text{distance moved by chemical}}{\text{distance moved by solvent}}$$

As long as the experiment is conducted under the same conditions (e.g. temperature, choice of mobile phase and stationary phase) then the R_f value of a substance will always be the same, and can often be found from a data book or reliable online source.

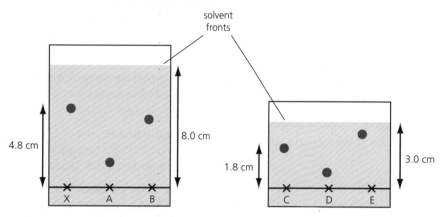

Figure 1.11 Calculating the R_f value for spot X on the first chromatogram gives an answer of 4.8 ÷ 8.0 = 0.6. Calculating the R_f value for spot C on the second chromatogram gives 1.8 ÷ 3.0 = 0.6. This suggests that X and C are the same substance

Required practical

Investigate paper chromatography using inks/food colourings

Method

1 A horizontal pencil line is drawn 1 cm up from the bottom of a rectangle of chromatography paper.
2 Small spots of two or more coloured liquids are placed onto the pencil line, spaced at least 2 cm apart, and allowed to dry.
3 A suitable solvent (e.g. water) is poured into a large beaker or chromatography tank to a depth of 0.5 cm.
4 The chromatography paper is carefully placed into the solvent so that it is supported from above and does not touch the sides of the container.
5 The chromatogram is removed from the solvent before the solvent reaches the top of the paper. The level of the solvent (the 'solvent front') is marked using a pencil.
6 The R_f values are calculated for each of the pigments.

Now test yourself

13 What method should be used to separate insoluble copper oxide from a solution of copper sulfate?
14 What is the name of the liquid which is produced during distillation?
15 Cherry cola is a mixture of water and several liquid flavourings. All of these substances have different boiling points. Suggest why simple distillation will not allow you to separate all of the liquids, and state the correct method needed.
16 During distillation, how should the cold-water supply be connected to the condenser?
17 What method should be used to obtain crystals of magnesium nitrate from a solution of magnesium nitrate?
18 During filtration, what name is used to describe the solid which is separated out?
19 In chromatography, what is meant by the term 'mobile phase'?
20 During paper chromatography when water is being used as the solvent, a brown ink has separated into three spots: red, green and blue. The red spot has not moved from the pencil line. What can you conclude about the red pigment?
21 Why can the R_f value of a substance never be more than 1?

Answers on page 123

Atomic structure

Discovering atomic structure

In the early 1800s, scientists thought that atoms were tiny indivisible spheres.

In the late 1800s, the **electron** was discovered by J.J. Thomson. He realised that it was a negatively charged particle which was part of an atom. The electron was therefore the first **sub-atomic particle** to be discovered, and this disproved the earlier model that atoms could not be split.

In the early 1900s, Ernest Rutherford interpreted the results of an experiment by Hans Geiger and Ernest Marsden to deduce that atoms had a central **nucleus** which contained the positive charge of the atom and almost all of the mass. The rest of the atom was mostly empty space, and the electrons were in this space, orbiting the nucleus.

Soon after, Niels Bohr suggested that electrons orbit in fixed shells.

By the mid-1900s, **protons** and **neutrons** had been identified in the nucleus of atoms.

The structure of a carbon atom is shown in Figure 1.12. The diagram is not to scale. In reality, the electrons would be orbiting much further from the nucleus (relatively speaking).

Electron: The sub-atomic particle which has a negative charge.

Sub-atomic particle: A tiny particle which is found inside atoms. There are three types of sub-atomic particle: protons, electrons and neutrons.

Nucleus: The cluster of protons and neutrons which is found at the centre of an atom.

Proton: The sub-atomic particle which has a positive charge.

Neutron: The sub-atomic particle which has no charge, so it is neutral.

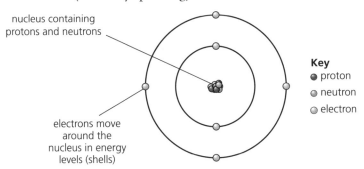

nucleus containing protons and neutrons

electrons move around the nucleus in energy levels (shells)

Key
● proton
◔ neutron
○ electron

Figure 1.12 A diagram of the structure of a carbon atom (not to scale)

Exam tip

You can use alliteration to help you remember the charges on the sub-atomic particles. **P**rotons are **p**ositive. **Neut**rons are **neut**ral. By elimination, electrons must be negative.

Protons, neutrons and electrons

Table 1.2 summarises the position, mass and electrical charge of protons, electrons and neutrons.

Table 1.2 Properties of the three sub-atomic particles

Particle	Position	Mass (relative to a proton)	Charge
proton	nucleus	1	+1
neutron	nucleus	1	0
electron	shells	1/1840	−1

Now test yourself

22 Which particles are found inside the nucleus of an atom?
23 What is the mass of a proton?
24 What is the charge on a neutron?

Answers on page 123

Atomic number and mass number

All atoms of an element have the same number of protons. This is called the **atomic number**. Atoms of different elements have a different number of protons. If you know the number of protons inside an atom, you can work out which element it is, because you can look up the proton number on a Periodic Table. In the Periodic Table, the atoms are arranged in order of increasing atomic number.

The **mass number** of an atom is the number of protons plus the number of neutrons. It is the total number of sub-atomic particles in the nucleus.

Atomic number: The number of protons in an atom.

Mass number: The total number of protons and neutrons in an atom.

Working out the number of protons, neutrons and electrons

Each element in the Periodic Table has a unique symbol, and this has two numbers alongside it. Use the key in your Periodic Table to check which is the atomic number and which is the mass number. In your exam, the mass number will be shown at the top of each element's box.

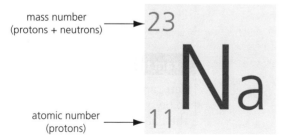

mass number (protons + neutrons) → 23

Na

atomic number (protons) → 11

Figure 1.13 The chemical symbol for sodium, including the mass number and atomic number

The atomic number and mass number can be used to work out the number of protons, neutrons and electrons in any atom.

Example

Using the sodium atom in Figure 1.13, we can work out the number of protons, neutrons and electrons:
- Atoms of sodium have 11 protons, because this is the atomic number.
- Atoms of sodium have 12 neutrons, because this is the mass number minus the atomic number.
- Atoms of sodium have 11 electrons, because every atom is neutral, so the number of electrons equals the number of protons.

Now test yourself

TESTED

25 What is the definition of the term 'atomic number'?
26 Describe how you can use the atomic number and mass number of an atom to calculate the number of neutrons in the atom.
27 Use a Periodic Table to find the atomic number of calcium.
28 Use a Periodic Table to calculate the number of protons, neutrons and electrons in an atom of lithium.
29 Use a Periodic Table to calculate the number of protons, neutrons and electrons in an atom of potassium.

Answers on page 123

Typical mistake

Students sometimes think that the number of neutrons is the mass number of the atom. Remember that for all of the atoms you will be asked about in the exam, the numbers of protons, electrons and neutrons will be similar to each other because the mass number is usually around twice the atomic number.

Revision activity

Choose a number at random between 1 and 20. This will be the atomic number of an element listed on the Periodic Table. Now calculate the number of protons, neutrons and electrons for an atom of your element.

Isotopes

What are isotopes?

REVISED

Isotopes are atoms of the same element which have different mass numbers. Isotopes must have the same number of protons as each other, as otherwise they would not be the same element. Isotopes have different masses because they have different numbers of neutrons.

Isotopes of an element have identical **chemical properties**. This is because the chemical reactions of an element depend on how easily it gains, loses or shares electrons. This depends on the number of protons and electrons in the atoms, and these are identical in all the isotopes.

Two isotopes of an element will have slightly different **physical properties**. For example, they will have very slightly different densities, and may have very slightly different melting and boiling points. Specific isotopes of some elements are radioactive and this makes them useful in medicine and in working out the age of the remains of living things that died a long time ago.

Isotopes: Atoms of the same element which have the same number of protons but different numbers of neutrons.

Chemical properties: Descriptions of the ways that a substance characteristically behaves in chemical reactions. For example, how reactive it is with oxygen, water or acids.

Physical properties: Descriptions of the ways that a substance characteristically behaves when it is not reacting with other chemicals. For example, its melting point, electrical conductivity, whether it is brittle, strong, hard, etc.

Protons, neutrons and electrons in isotopes

REVISED

There are two isotopes of bromine: $^{79}_{35}$Br and $^{81}_{35}$Br. We can calculate the number of protons, neutrons and electrons in each of these two isotopes, and they are summarised in Table 1.3.

Table 1.3 A comparison of the sub-atomic particles in two isotopes of bromine.

	$^{79}_{35}$Br	$^{81}_{35}$Br
Protons	35	35
Neutrons	44	46
Electrons	35	35

Revision activity

Summarise your revision of isotopes in a table with two headings: similarities and differences. Include in your table a comparison of the number of each sub-atomic particle, and also the physical and chemical properties.

Relative atomic mass

REVISED

The average mass of a bromine atom is what is provided on the Periodic Table. In the case of bromine, half of all bromine atoms have a mass number of 79 and the other half have a mass number of 81, so the average mass, called the **relative atomic mass (A_r)**, for bromine is 80.

For most elements, the isotopes are not present in equal proportions. This means that the average mass of an atom of that element will not be halfway between the mass numbers of the isotopes.

The most (average) mass of an atom of an element, taking into account the **isotopic abundance**, is known as the relative atomic mass (A_r) of that element. The relative atomic mass is always compared with carbon, on a scale where the mass of a carbon-12 atom is exactly 12.

Chlorine has two isotopes: $^{35}_{17}$Cl and $^{37}_{17}$Cl. Both isotopes of chlorine have 17 protons and 17 electrons in every atom. However, chlorine-35 has 18 neutrons in each atom, while chlorine-37 has 20 neutrons in every atom.

In a naturally occurring sample of 100 chlorine atoms, 75 of them are ^{35}Cl and 25 are ^{37}Cl. This skews the average away from the halfway point (36) and makes the relative atomic mass of chlorine less than 36.

The relative atomic mass of an element with two isotopes is calculated using the equation below.

Relative atomic mass =

$$\frac{(\text{mass of isotope 1} \times \% \text{ of isotope 1}) + (\text{mass of isotope 2} \times \% \text{ of isotope 2})}{100}$$

Relative atomic mass (A_r): The mean mass of an atom of an element, taking into account the relative proportions of naturally occurring isotopes, relative to 1/12 the mass of a carbon-12 atom.

Isotopic abundance: The percentage of a particular isotope in a sample of atoms of that element. The sum of the abundances of all the isotopes will be 100.

Exam tip

When you calculate the relative atomic mass of an element, remember that you are calculating an average of the masses of the particles. Therefore, your final answer needs to be somewhere in between the heaviest and lightest isotopes, e.g. for chlorine, if your answer was 41, you would know you had made a mistake as it must be between 35 and 37.

Example

Using chlorine as an example:

- Relative atomic mass = $\dfrac{(35 \times 75) + (37 \times 25)}{100}$

  $= \dfrac{2625 + 925}{100}$

  $= 35.5$

30 Calculate the number of protons, neutrons and electrons for these two atoms, and explain why they are isotopes: $^{12}_{6}C$ and $^{14}_{6}C$.

31 Why would it be hard to distinguish between samples of chlorine-35 and chlorine-37 in the laboratory?

32 Boron has two isotopes: ^{10}B and ^{11}B. In a sample of boron atoms, 20% are boron-10 and the remaining 80% are boron-11. Calculate the relative atomic mass of boron.

Answers on page 123

The arrangement of electrons

Groups and periods

REVISED

The elements in the Periodic Table are arranged in horizontal rows called **periods** and vertical columns called **groups**. The first period contains only two elements: hydrogen and helium. Period 2 contains the elements lithium to neon.

Groups contain elements with similar chemical properties. For example, all of the elements in group 1 react in similar ways with water. The elements in the middle of the table are called the **transition metals** and they are not usually given group numbers. The elements in the final column are labelled as group 0 (not group 8 as you might expect), and they all have similar properties.

Period: A horizontal row of elements in the Periodic Table.

Group: A vertical column of elements in the Periodic Table, which have similar chemical properties.

Transition metals: The metals which are found in the Periodic Table between group 2 and group 3 (see page 20).

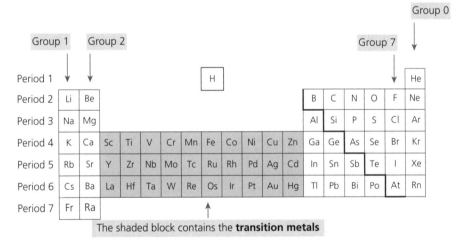

Figure 1.14 A simplified Periodic Table with periods and some groups labelled, and the transition metals shaded

Electron configurations of atoms

REVISED

Remember that the number of electrons in an atom is equal to the atomic number of that element. Electrons orbit the nucleus in shells, or energy levels. The shell closest to the nucleus has the lowest energy, so it is filled first. When the first shell is filled, electrons begin to fill the second shell, and then the third, and so on. You need to know that:

● The first shell can hold a maximum of two electrons.
● The second shell can hold a maximum of eight electrons.
● The third shell can hold a maximum of eight electrons.
● Any remaining electrons go into the fourth shell.

This model works for elements up to calcium. If you study chemistry beyond GCSE you will be taught a more advanced model that works for heavier atoms.

If we take sulfur as an example, to work out the electron configuration you would:

- See that the atomic number of sulfur is 16, so there are 16 protons in every atom of sulfur.
- This means there must also be 16 electrons, to make the atom neutral.
- Two electrons can fit into the first shell, then it is full so you need to move onto the next shell.
- Eight further electrons can fit into the second shell, then it is full. You have now used up ten electrons, and have six spare.
- The six remaining electrons go into the third shell.
- The electron configuration is represented like this: 2,8,6.

When the electron configuration diagrams are arranged in the positions of the elements in the Periodic Table, it is easy to see two important trends:

- The period number of an element is the number of shells that are filled or partially filled with electrons in atoms of that element.
- The group number of an element is the number of electrons in the outer shell of atoms of that element.

For example:

- Boron is in period 2 and group 3. Its electron configuration is 2,3. It has two shells, and three electrons in its outer shell.
- Chlorine is in period 3 and group 7. Its electron configuration is 2,8,7. It has three shells and seven electrons in its outer shell.
- Potassium is in period 4 and group 1. Its electron configuration is 2,8,8,1. It has four shells and one electron in its outer shell.

Important things to note are that:

- Hydrogen has one electron (in its outer shell) but it is not placed in group 1 because it does not react in a similar way to the elements that are in group 1.
- Helium has two electrons in its outer shell, but the other elements in group 0 have eight electrons in their outer shell.
- The elements in group 0 all have full outer shells, and this is what causes them to have similar properties.

> **Exam tip**
>
> It is recommended that you arrange the electrons in the second, third and fourth shells at the north, east, south and west positions, in pairs. This makes it much easier for you (and the examiner) to count them, and this means that you are less likely to make mistakes.

Important groups in the Periodic Table

REVISED ☐

Group 1: the alkali metals

The elements in group 1 are called the **alkali metals**. You will find more detail on them on page 50.

- The atoms of the elements in group 1 all have one electron in their outer shell.
- When they react with other substances, these atoms all lose their one outer electron to form a stable **ion** with an empty outer shell.
- When these atoms lose one electron, they form a stable ion with a single positive charge, e.g. Li^+ or Na^+.

lithium
Li
2, 1

sodium
Na
2, 8, 1

Figure 1.15 Electron configurations of the first two elements in group 1

Exam practice answers and quick quizzes at **www.hoddereducation.co.uk/myrevisionnotesdownloads**

Group 7: the halogens

The elements in group 7 are called the **halogens**. You will find more detail on them on page 52.

- The atoms of the elements in group 7 all have seven electrons in their outer shell.
- When they react with other substances, these atoms all gain or share one electron to form a stable ion with a full outer shell.
- When these atoms gain one electron, they form a stable ion with a single negative charge, e.g. F^- or Cl^-.

fluorine
F
2, 7

chlorine
Cl
2, 8, 7

Figure 1.16 Electron configurations of the first two elements in group 7

Alkali metals: The elements which are found in group 1 of the Periodic Table.

Ion: A charged particle which is formed from an atom (or group of atoms) that has gained or lost electrons.

Halogens: The elements which are found in group 7 of the Periodic Table.

Exam tip

Remember that to explain why elements in the same group have similar chemical properties, you need to state that they have the same number of electrons in their outer shell. This means they will gain, lose or share the same number of electrons as each other.

Exam tip

Remember that electrons are negative, so when a non-metal atom gains one or more electrons, it forms a negative ion. When a metal loses one or more electrons, it forms a positive ion.

Forming stable ions

REVISED

In chemical reactions, atoms of elements tend to gain, lose or share electrons to achieve a stable outer shell which is either full or empty. The group number tells us the number of electrons in the outer shell, and this allows us to predict what change is needed for that atom to obtain a full or empty outer shell.

Example

Work out the charge on an ion of a metal in group 2.
- They have two electrons in their outer shell.
- They can achieve a stable empty outer shell by losing two electrons.
- This loss of two electrons means they form an ion with a charge of 2+.
- Examples of this include Mg^{2+} and Ca^{2+}.

Example

Work out the charge on an ion of a non-metal in group 6.
- They have six electrons in their outer shell.
- They can achieve a stable empty outer shell by gaining two electrons.
- They gain two electrons to form an ion with a charge of 2–.
- Examples of this include O^{2-} and S^{2-}.

Atoms of the elements in Group 0 have a full outer shell, which is stable, so they do not need to gain, lose or share any electrons. This is why they do not react with other chemicals.

Now test yourself

33 How many electrons can fit into the first and second electron shells?
34 Atoms of phosphorus have 15 electrons. Deduce the electron configuration of phosphorus.
35 Use the Periodic Table to identify the element whose atoms have the electron configuration 2,8,3.
36 What is the electron configuration of a fluoride ion, F⁻?
37 Work out the number of protons, neutrons and electrons, and the electron configuration in a sodium atom. Repeat this for a stable sodium ion, which has an empty outer shell.

Answers on page 123

> **Exam tip**
>
> Metals always react by losing electrons to form positive ions. Non-metals in groups 6 and 7 can either gain electrons to form negative ions, or share electrons to achieve a full outer shell. Non-metals in groups 4 and 5 do not normally form ions, and almost all of their compounds are therefore covalent (which means they involve shared electrons).

Measuring atoms

Atoms are very small indeed and have a typical diameter of 0.1 nanometres, which is 1×10^{-10} metres. An atom has a tiny mass as well, which means we have to use a relative scale, i.e. comparing the weight of an atom of an element to one-twelfth the mass of a carbon-12 atom. This is similar to the mass of a hydrogen atom. This method of comparing the masses of atoms tells us that a magnesium atom, ^{24}Mg, has a mass which is twice that of a carbon atom, ^{12}C.

The relative atomic mass of an element is the mean mass of an atom of that element, taking into account the relative proportions of naturally occurring isotopes, relative to 1/12 the mass of a carbon-12 atom. The relative atomic mass of an element is given the symbol A_r.

Now test yourself

38 Which is larger, the nucleus of an atom or the nucleus of a plant cell?
39 Which is the standard element to which the masses of other atoms are compared?
40 What is the symbol for relative atomic mass?

Answers on page 123

Patterns of elements

Early attempts at classifying the elements

In the early 1800s, Johann Dobereiner grouped elements into **triads** (groups of three) if they had similar chemical properties.

In the mid-1800s, John Newlands arranged the elements which had been discovered by this time in order of increasing atomic mass. Newlands proposed his '**Law of Octaves**' which stated that every eighth element had similar properties to the element which was seven places earlier. Unfortunately, there were some elements that did not fit the pattern. Newlands was laughed at by other scientists, who did not recognise the usefulness of his ideas. The idea of a repeating pattern of chemical properties is very important, and we now call this '**periodicity**'.

> **Periodicity**: The repeating pattern of chemical properties when the elements are arranged into the Periodic Table. This means that an element is likely to be very similar to the elements above and below it in the same group.

Mendeleev's Periodic Table

In the late 1800s, Dmitri Mendeleev published his Periodic Table. He also arranged elements in order of increasing atomic mass, but he did two very clever things in addition:

● He swapped the positions of some elements so that their chemical properties fitted more neatly into the repeating pattern of properties.
● He left some gaps in order to allow some elements to fit more neatly into the repeating pattern of properties. He predicted that these gaps would later be filled with newly discovered elements, and he even successfully predicted the properties of those undiscovered elements.

Part of Mendeleev's Periodic Table is shown in Figure 1.17. The asterisks (*) show the positions of some of the missing elements that Mendeleev predicted. Notice that elements with similar chemical properties fit neatly into vertical columns (groups), for example lithium, sodium and potassium. Also notice that there are no noble gas elements (now placed in group 0) because they had not been discovered at that time.

Mendeleev's Periodic Table was widely accepted by other scientists because it helped to explain the patterns in the chemical reactions of the elements, and because he correctly predicted the properties of undiscovered elements.

Revision activity

Summarise the information in this section into a timeline sequence. You don't need to memorise the dates of the important developments.

	I	II	III	IV	V	VI	VII	VIII
Period 1	H							
Period 2	Li	Be	B	C	N	O	F	
Period 3	Na	Mg	Al	Si	P	S	Cl	
Period 4	K / Cu	Ca / Zn	* / *	Ti / *	V / As	Cr / Se	Mn / Br	Fe Co Ni
Period 5	Rb / Ag	Sr / Cd	Y / In	Zr / Sn	Nb / Sb	Mo / Te	* / I	Ru Rh Pd

GROUP

Figure 1.17 Part of Mendeleev's Periodic Table

Now test yourself

TESTED

41 What term is given to the repeating pattern of the properties of the chemical elements?
42 Which scientist proposed the Law of Octaves?
43 Which group of elements found on the modern Periodic Table was missing from Mendeleev's version?

Answers on pages 123–4

The modern Periodic Table

Developments after Mendeleev

One of the reasons why Mendeleev's table was such a successful tool in chemistry was that it helped scientists make predictions about elements which had not been discovered yet. Another reason was that his Periodic Table could be improved as new evidence became available.

Helium was discovered in 1895 and could not fit into the Periodic Table because it was not like any other element. A new group was needed, and chemists then focused on finding other similar elements to go into this group.

The **transition metals** were moved to the middle of the table so that the elements could be arranged in order of increasing atomic number instead of increasing atomic mass.

> **Transition metals** (also called transition elements): The elements which are located between groups 2 and 3 in the Periodic Table. They mostly have similar physical and chemical properties.

Key facts about the modern Periodic Table

REVISED

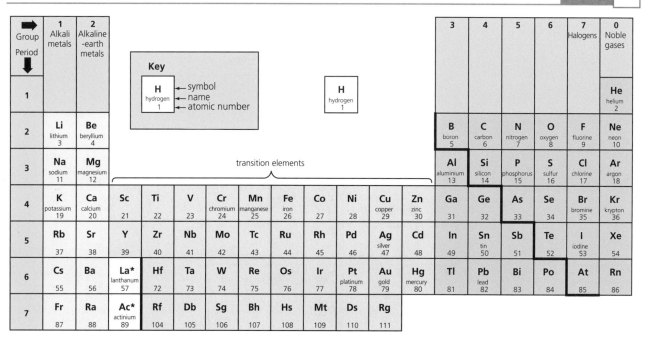

Figure 1.18 The modern Periodic Table, omitting the elements with atomic numbers 58–71 and 90–103

- The 'staircase line' separates metals from non-metals. Metals are on the left and in the middle of the table. Non-metals are on the right. Remember that hydrogen is also a non-metal. Also remember that some elements very close to the staircase line are semi-metallic.
- The horizontal rows are called periods and these show the number of electron shells that are filled or partially filled in atoms of that element.
- The vertical columns are called groups and these show the number of electrons in the outer shell of atoms of that element.
- The groups contain elements that have similar chemical properties and that show trends in their physical properties.

> **Exam tip**
>
> Remembering that B is a non-metal and Al is a metal will help you to write in the staircase line separating metals from non-metals for yourself in the exam.

TESTED ☐

Now test yourself

44 What name is given to a horizontal row in the Periodic Table?
45 What is the name given to the elements in group 7?
46 In which group do you find the alkali metals?
47 Which element is in period 4, group 1?

Answers on page 124

The noble gases

Discovering the noble gases

REVISED ☐

The noble gases were discovered relatively recently in the history of chemistry because they are unreactive and therefore it is hard to detect them using chemical reactions.

Helium is named after the Sun because it was discovered as a new element by scientists who were observing light from the Sun. The wavelength of the light could not have come from any element that had yet been discovered on Earth. After searching for it, helium was later discovered in some rocks and natural gas reserves.

Argon was discovered by scientists who were investigating the gases present in air. When they accounted for the nitrogen, oxygen and carbon dioxide, there was another mystery gas present which made up almost 1% – this was argon.

After group 0 had been added to the Periodic Table, the remaining noble gases were discovered quite quickly.

Properties of the noble gases

REVISED ☐

The elements in group 0 are called the noble gases because they are **inert**. This means that they do not normally react with other elements. The elements do not react because they all have a full outer shell of electrons, which is stable. This means that they do not need to gain, lose or share electrons from other atoms. It also means that the noble gases are all made of single atoms, not molecules like other gases. Therefore, helium is He, not He_2, neon is Ne, not Ne_2, and so on. We say that the noble gases are **monatomic**.

As you go down group 0, there are trends in the density and boiling point of the noble gases. Both of these properties increase down the group, as shown by the graphs in Figure 1.20, on the following page.

> **Inert**: A substance that does not react with other chemicals.
>
> **Monatomic**: An element which is made of individual atoms which are not chemically bonded to each other.

 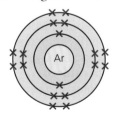

Figure 1.19 The electron configurations of two elements in group 0, showing their full outer shells

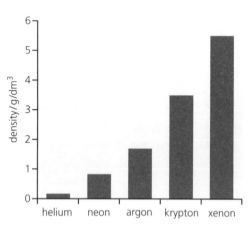
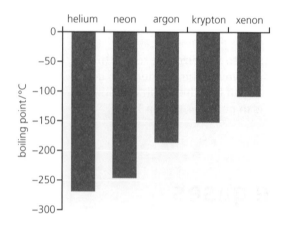

Figure 1.20 The densities (left) and boiling points (right) of the elements in group 0

Uses of the noble gases

The typical uses of the noble gases are shown in Table 1.4. They can be explained by thinking carefully about their properties.

Table 1.4 The typical uses of some of the noble gases

Noble gas	Used for	Relevant property
Helium	Party balloons and weather balloons	Very low density, so the balloons float in air
Neon	Red-coloured advertising lights	Produces coloured light when electricity is passed through it
Argon	Welding, to prevent hot metal reacting with oxygen	Inert
Krypton	Lasers	Produces intense light when electricity is passed through it

Revision activity

Copy out Table 1.4 with the key information missing. After revising another section, see if you can fill in the table from memory. Check your answers to see if you are correct.

Now test yourself

48 Why are the noble gases inert?
49 Which noble gas has the highest density?
50 What is the trend in boiling point as you go down group 0?
51 Which element is used in weather balloons that are used to carry atmospheric sensors high into the atmosphere?

Answers on page 124

Summary

- In a solid, the particles are fixed in place and vibrating.
- In a liquid, the particles are mostly touching their neighbours and can move over each other.
- In a gas, the particles are widely spaced and moving quickly.
- Diffusion occurs rapidly in gases (and slowly in liquids) because the particles of one substance are knocked around by the particles of another substance, and then randomly spread out from an area of high concentration.

- An impure substance will melt or boil at a temperature different from the expected temperature, and it may melt or boil over a range of temperatures.
- Filtration separates an insoluble solid from a liquid.
- Crystallisation separates a solid solute from a solution.
- Simple distillation separates a solvent from a solution.
- Fractional distillation separates a mixture of liquids on the basis that they have different boiling points.

➡

- Paper chromatography separates substances in a mixture depending on their different attraction to the mobile phase (solvent) and the stationary phase (paper).
- The R_f value is calculated by dividing the distance a substance moves by the distance the solvent moves. R_f values are always decimals less than 1.
- Protons are positive and have a relative mass of 1.
- Electrons are negative and have almost no mass.
- Neutrons are neutral and have a relative mass of 1.
- When an atom gains electrons, it becomes a negative ion.
- When an atom loses electrons, it becomes a positive ion.
- The atomic number is the number of protons in an atom.
- The number of electrons in an atom is the same as the number of protons.
- The number of neutrons in an atom is the mass number minus the atomic number.
- Isotopes have the same number of protons, but different numbers of neutrons.
- The relative atomic mass of an element is the mean mass of an atom of an element, taking into account the relative proportions of naturally occurring isotopes, relative to 1/12 the mass of a carbon-12 atom.

- In the Periodic Table, periods are the horizontal rows and groups are the vertical columns.
- Groups contain elements with similar chemical properties and trends in physical properties.
- The first (innermost) electron shell can hold a maximum of two electrons, the second shell can hold up to eight electrons, and the third shell can hold up to eight electrons.
- The number of full or partially-filled shells in an atom is the same as the period number of that element.
- The number of electrons in the outer shell of an atom is the same as the group number of that element. However, elements in group 0 all have full outer shells, and this is a stable electron configuration.
- Mendeleev's Periodic Table was an improvement on earlier versions based solely on atomic mass because he swapped the position of some elements in order to allow them to fit into groups with similar elements, and he left gaps for undiscovered elements.
- Elements are now arranged in order of increasing atomic number. Metals are on the left and in the middle of the Periodic Table. To the right of the 'staircase line' are the non-metals.
- The noble gases are inert (unreactive) because they have full outer shells of electrons, so they do not need to gain, lose or share electrons.

Exam practice

1 Drinking water is not readily accessible for some people in the world. Instead they must either drink dirty water or find a way to purify the water they have available.

a Sea water is not safe to drink because it contains a number of dissolved compounds. The most abundant of the dissolved compounds in sea water is sodium chloride. Use terms from the list below to complete a copy of the table, which relates to sea water. [3]

solution   **solute**   **solvent**

Sodium chloride is a...	
Water is a...	
The mixture of sodium chloride and water is a...	

b It is possible to distil sea water to obtain water that is safe to drink. Describe how to distil sea water, using a labelled diagram in your answer. [5]

c Pure water boils at exactly 100 °C. A pair of students decided to test a bottle of mineral water with the word 'pure' on the label.
They took a sample of water from the bottle and measured its boiling point. It began to boil at 100.7 °C, but the temperature of the boiling water increased to 101.3 °C. This was the highest temperature recorded. Explain why the boiling point and temperatures were not what the students expected. [3]

2 Atoms are the smallest particles of an element.

a What is the central part of an atom called? [1]

b Complete a copy of the table to show the mass and charge of the three types of sub-atomic particle. [6]

Sub-atomic particle	Mass	Charge
Proton		
Electron		
Neutron		

c What is meant by the following terms?
  i   Atomic number [1]
  ii Mass number [1]
d Name the element that has three electron shells, and has one electron in its outer shell.
   Use the Periodic Table to help you. [1]
e Copper has two isotopes, ^{63}Cu and ^{64}Cu. In a sample of copper, the proportion of ^{63}Cu is 70%,
   with the remaining 30% being ^{64}Cu. Calculate the relative atomic mass of copper in this sample. [3]

3 Chromatography can be used to separate pigments that are mixed together in an ink. Look at the
  chromatogram on the right. 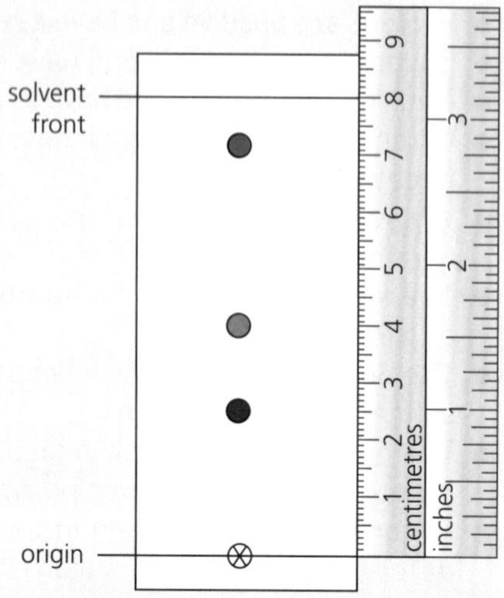
  a State why the start line must be drawn in pencil and
     not pen. [1]
  b Identify how many pigments were in this mixture. [1]
  c Identify which coloured pigment had the greatest attraction
     for the mobile phase. [1]
  d Calculate the R_f value of the green pigment. [2]
  e Explain why a substance might have an R_f value of zero. [2]

4 The solubility of a substance is the mass of it that can be
  dissolved in a specific mass (usually 100 g) of water.
  a Describe how you would measure the solubility of sodium
     nitrate at 50 °C, using the units g of solute per 100 g of water. [6]
  b The graph shows how the solubilities of three compounds
     change with temperature.

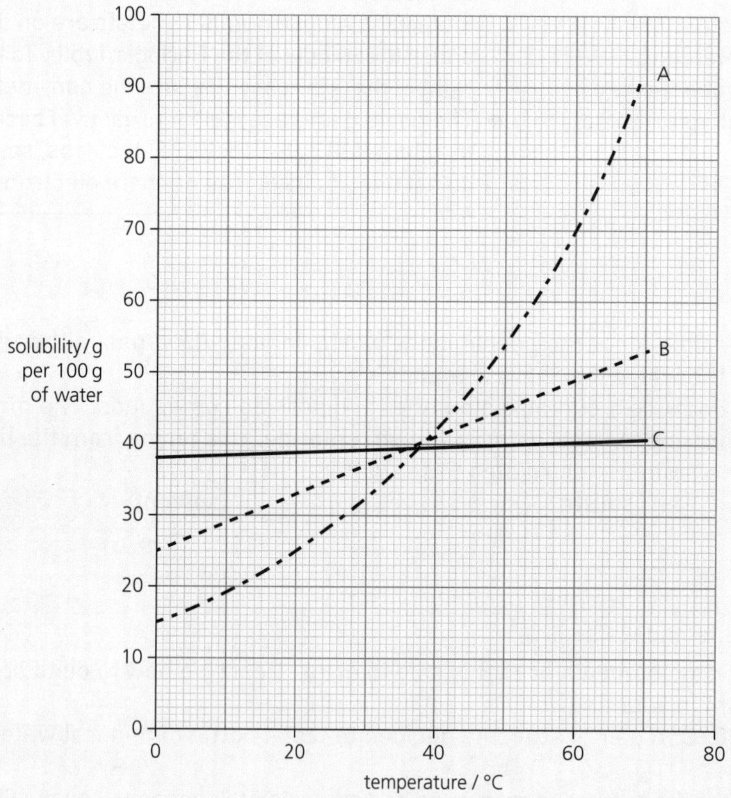

  i   State the solubility of substance A at 60 °C. [1]
  ii   Identify which substance is the most soluble at 30 °C. [1]
  iii Calculate what mass of substance B can be dissolved in 50 g of water at 20 °C. [2]
  iv   A saturated solution of A is made at 50 °C using 100 g of water. It is then allowed to
       cool to 20 °C, and some crystals form. Calculate the theoretical mass of the crystals
       of A that have formed. [3]

Answers and quick quizzes online

ONLINE

1 Principles of chemistry 2

Using relative atomic masses

Relative formula mass

The **relative atomic mass**, A_r, is the average mass of an atom of an element on a scale where the mass of a carbon-12 atom is exactly 12. The **relative formula mass**, M_r, of a molecular substance is the mass of one molecule on the same scale, so it is calculated by adding up the A_r values for all the atoms in a molecule.

> **Relative atomic mass:** The mean (or average) mass of an atom of an element, taking into account the relative proportions of naturally occurring isotopes (see page 13), relative to 1/12 the mass of a carbon-12 atom.

> **Example**
>
> To calculate the M_r of carbon dioxide (CO_2), you add up the total of the A_r values for one carbon and two oxygen atoms. So $12 + 16 + 16 = 44$.

You can also calculate the M_r for a molecular element or an ionic compound in the same way, even though ionic compounds are not made of molecules.

> **Relative formula mass:** The sum of the relative atomic masses in one 'formula unit' of a substance, whether this is a molecular element or a compound.

> **Example**
>
> To calculate the M_r of nitrogen (N_2), you add up the total of the A_r values for two nitrogen atoms. So $14 + 14 = 28$.

> **Example**
>
> Calculate the M_r of ammonium sulfate, $(NH_4)_2SO_4$.
>
> - From the Periodic Table we can see: $^{14}_7N$, ^1_1H, $^{32}_{16}S$ and $^{16}_8O$.
> - In the formula, we can see that there are **two N atoms**, **eight H atoms**, **one S atom** and **four O atoms**. Remember to multiply the numbers of N and H atoms inside the brackets by the subscripted number 2 after the brackets.
> - So the $M_r = (2 \times 14) + (8 \times 1) + 32 + (4 \times 16)$
> $= (28) + (8) + 32 + (64) = 132$

> **Exam tip**
>
> Where a formula contains brackets, remember to multiply the number of each type of atom inside the brackets by the subscripted number after the brackets to correctly calculate the number of each type of atom **first**. Then work out the M_r second.

The number of atoms in a mole

Some people might count eggs in dozens. A dozen eggs is 12 eggs. Two dozen eggs is 24 eggs. This is similar to how chemists count atoms, in that they do not count them one by one. The name of the quantity chemists use is not a dozen – it is called a **mole** (abbreviated to mol). The number it represents is not 12 but instead it is a very large number, called the **Avogadro number** (named after an Italian scientist), and it is 6×10^{23}. One mole of atoms of an element contains 6×10^{23} atoms.

> **Mole:** The quantity used to measure the amount of substance in chemistry. One mole of any atomic element always contains the same number of atoms.

> **Avogadro number:** The number of particles in a mole of a substance. The Avogadro number is 6×10^{23}.

Example

Calculate the number of atoms in three moles of aluminium.

- Remember that one mole of any element contains 6×10^{23} atoms.
- Therefore three moles of any element contains $3 \times (6 \times 10^{23}) = 1.8 \times 10^{24}$ atoms.

Exam tip

Calculations involving the Avogadro number give very large answers indeed. You must practise using your scientific calculator so that you know how to enter and interpret numbers which are in standard form.

The mass of a mole

REVISED

One mole of an element which is made of atoms has a mass in grams which is equal to the relative atomic mass of that element. For example, one mole of carbon atoms (^{12}C) has a mass of 12 g.

For an element that is made of molecules (e.g. O_2), one mole has a mass equal to the relative formula mass, M_r, of that substance. For example, one mole of $^{16}O_2$ molecules has a mass of 32 g because the M_r of an oxygen molecule is 32. This also works for any compound: one mole is equal to the M_r in grams.

Example

What is the mass of one mole of iron?

- The relative atomic mass of iron can be found from the Periodic Table. It is 56.
- Therefore, the mass of one mole of iron is 56 g.

Example

Calculate the mass of one mole of sodium chloride, NaCl.

- Work out the M_r by adding up the total of the A_r values in the formula: $23 + 35.5 = 58.5$
- Remember that one mole will have a mass equal to the M_r, so one mole is 58.5 g.

Chemists use an equation to calculate the number of moles present in a given mass of a substance:

$$\text{number of moles (mol)} = \frac{\text{mass of sample}}{M_r}$$

If the substance is an element, the A_r should be used instead of the M_r.

Example

Calculate the number of moles in 400 g of $CaCO_3$.

- Calculate the M_r of $CaCO_3$: $40 + 12 + (3 \times 16) = 100$
- Substitute the numbers into the equation:

$$\text{number of moles (mol)} = \frac{400}{100} = 4 \text{ mol}$$

The moles equation can also be rearranged to work out the mass of a specified number of moles of a given substance:

$$\text{mass (g)} = \text{number of moles} \times M_r$$

Example

Calculate the mass of two moles of NaCl.

- Work out the M_r of NaCl: $23 + 35.5 = 58.5$
- Substitute the numbers into the equation:

$$\text{mass (g)} = 2 \times 58.5 = 117 \text{ g}$$

Exam tip

Practise rearranging this equation so that you are confident when it comes to your exam because you will not be provided with it. It is worth learning the two versions of the equation which are given in these examples, and you are most likely to need to calculate the moles or the mass, as opposed to the M_r.

Moles in equations

When equations are balanced (see pages 31–2), the numbers in front of the formulae represent the number of moles of each of the reactants and products. For example, in the following equation:

$$2Na + Cl_2 \rightarrow 2NaCl$$

we can see that two moles of sodium react with one mole of chlorine molecules to produce two moles of sodium chloride. Note that the number of moles at the start (three in this example) does not need to equal the number of moles at the end of the reaction (two in this case).

Calculating a chemical formula using experimental data

A chemical formula which is calculated from the masses of reacting substances measured in an experiment is called an **empirical formula**. It is the simplest ratio of atoms of each element in a compound. For example, the compound ethane has the **molecular formula** C_2H_6, but a ratio of 2:6 can be simplified to 1:3, so the empirical formula of ethane is CH_3.

> **Empirical formula:** The simplest ratio of the atoms of each element in a compound.
>
> **Molecular formula:** The actual ratio of the atoms of each element in a molecule of a compound.

Working out the empirical formula of water

When 36 g of water are decomposed into hydrogen and oxygen, 32 g of oxygen is produced, with the remaining 4 g being hydrogen. The empirical formula of water can be calculated using the following steps.

Table 1.5

	Hydrogen	Oxygen
Reacting masses (g)	4	32
A_r	1	16
Moles of atoms = mass ÷ A_r	4 ÷ 1 = 4	32 ÷ 16 = 2
Simplified ratio	2	1
Empirical formula	H_2O	

Required practical

Determine the formula of a metal oxide by combustion (e.g. magnesium oxide)

Method

The starting mass of the metal needs to be subtracted from the final mass of the metal oxide to deduce the mass of the oxygen atoms that reacted.

1 The mass of a clean dry crucible (with lid) was measured using an accurate top pan balance.
2 A piece of magnesium ribbon was placed into the crucible and it was weighed again.
3 The mass of magnesium was calculated by subtracting the weight of the crucible from the weight of the crucible plus the magnesium.
4 The crucible and lid were placed on a pipe clay triangle over a Bunsen burner and heated strongly. →

5 From time to time, the lid was lifted with tongs to allow oxygen atoms from the air to react with the magnesium atoms. N.B. If the lid was lifted too often, the solid white magnesium oxide product would have been lost as smoke.

6 After ten minutes of heating, the crucible, lid and contents were weighed again.

7 The crucible, contents and lid were heated for another two minutes, then weighed again.

8 If the mass was found to be constant, the experiment was ended. If the mass was found to have increased, step 7 was repeated until a constant mass was achieved, which confirmed that the oxidation reaction was complete.

Analysis

- Mass of crucible and lid = 50.0 g
- Mass of crucible, lid and Mg = 51.2 g
- Mass of Mg at start: 51.2 − 50.0 = 1.2 g
- Maximum mass of crucible, lid and MgO after reaction has finished: 52.0 g
- Mass of magnesium oxide produced: 52.0 − 50.0 g = 2.0 g
- Mass of oxygen atoms that reacted: 2.0 − 1.2 = 0.8 g

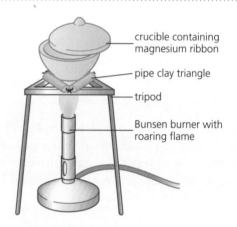

Figure 1.21 **The apparatus used to determine the empirical formula of magnesium oxide**

crucible containing magnesium ribbon
pipe clay triangle
tripod
Bunsen burner with roaring flame

	Magnesium	Oxygen
Reacting masses (g)	1.2	0.8
A_r	24	16
Moles of atoms = mass ÷ A_r	1.2 ÷ 24 = 0.05	0.8 ÷ 16 = 0.05
Simplified ratio	1	1
Empirical formula	MgO	

Working out the formula of a hydrated salt

Some ionic compounds have a specific number of molecules of water trapped within the giant ionic lattice structure. This gives them a formula which looks like this: $CuSO_4\ 5H_2O$. It is unusual to see a number which is not subscripted within a chemical formula. This formula is interpreted as follows: there are five water molecules for every copper ion and sulfate ion. You can work out the number of moles of water in a mole of hydrated salt by using an approach similar to the empirical formula calculations: a weighed sample of the hydrated salt is heated to constant mass which drives off all the water.

Example

27.8 g of hydrated iron(II) sulfate was heated to a constant mass. The mass of the dehydrated iron(II) sulfate was 15.2 g. Calculate the mass of water which was lost, and then the formula of the hydrated iron(II) sulfate.
- Calculate the mass of water lost: 27.8 − 15.2 = 12.6 g
- Now calculate the numbers of moles of salt and water present, then simplify the ratio.

	$FeSO_4$	H_2O
Reacting masses (g)	15.2	12.6
M_r	152	18
Moles = mass ÷ M_r	15.2 ÷ 152 = 0.1	12.6 ÷ 18 = 0.7
Simplified ratio	1	7
Empirical formula	$FeSO_4.7H_2O$	

Now test yourself

1 Use the Periodic Table to find the A_r of aluminium.
2 How many atoms are there in two moles of Al?
3 Calculate the M_r of aluminium oxide, Al_2O_3.
4 What is the mass of one mole of Al_2O_3?
5 How many moles are there in 204 g of Al_2O_3?
6 31.75 g of copper was heated in air. Its mass increased to 39.75 g. Calculate the empirical formula of copper oxide.

Answers on page 124

TESTED

Writing formulae

Compounds of non-metals with non-metals

There are a number of formulae that you are likely to know by heart by the time you take your exam. Examples include H_2O, CO_2, and the diatomic elements H_2, O_2, N_2, Cl_2, etc. To deduce the formula of a **covalent** compound, you need to know how many bonds each atom is likely to form. This can be estimated by the position of the element in the Periodic Table. It is also essential to remember that hydrogen always forms one covalent bond.

> **Covalent:** A compound that is made from non-metal elements. Covalent bonds are pairs of electrons which are shared between two atoms. Most covalent compounds are made of molecules.

Table 1.6 The number of covalent bonds formed by some non-metal atoms

Non-metal in group...	Examples	Likely to form this many bonds
4	Carbon, silicon	4
5	Nitrogen	3
6	Oxygen, sulfur	2
7	Fluorine, chlorine, bromine, iodine	1

> **Example**
>
> The compound ammonia is made from nitrogen and hydrogen. Deduce its formula.
> - Nitrogen forms 3 bonds.
> - Hydrogen forms 1 bond.
> - You will need three hydrogen atoms to satisfy the requirement of the nitrogen.
> - Therefore, the formula is NH_3.

Compounds of metals and non-metals

Compounds which contain metal atoms and non-metal atoms are **ionic**. This means that they are made from positive and negative **ions** which are strongly attracted in a giant structure (see page 38).

Metal atoms form positive ions by losing the electrons from their outer shells (see pages 16–17). Since the number of electrons in their outer shell is the same as their group number, it is easy to predict their ionic charge if they are in group 1, 2 or 3.

Non-metal atoms form negative ions by gaining electrons to fill their outer shells (see pages 16–17). It is also easy to predict their ionic charge if they are in group 6 or 7.

> **Ionic:** A compound that is made from a metal and one or more non-metal elements.
>
> **Ion:** A tiny particle with a positive or negative electrical charge. Ions are formed when atoms or groups of atoms gain or lose electrons.

Table 1.7 The charges on some common ions, linked to their position on the Periodic Table

Group 1	Group 2	Group 3	Transition metals (examples)	Group 6	Group 7
Li^+	Be^{2+}		Cu^{2+}, Fe^{2+}, Fe^{3+}, Ag^+, Zn^{2+}	O^{2-}	F^-
Na^+	Mg^{2+}	Al^{3+}		S^{2-}	Cl^-
K^+	Ca^{2+}				Br^-
					I^-

Exam tip

Transition metals have their ionic charge specified in their names, e.g. iron(III) is Fe^{3+} and copper(II) is Cu^{2+}. There are some complex ions whose formulae you need to learn for the exam, which are listed in Table 1.8.

Table 1.8 The ionic charges on some complex ions

Positive ions		Negative ions	
Ammonium	NH_4^+	Hydroxide	OH^-
		Sulfate	SO_4^{2-}
		Nitrate	NO_3^-
		Carbonate	CO_3^{2-}

Typical mistake

Do not confuse the sulf**ide** ion (S^{2-}) with the sulf**ate** ion (SO_4^{2-}). The sulfate ion is much more common. A good way to remember is that when a negative ion ends in the letters 'ate' you know it contains oxygen atoms.

To work out the formula of an ionic compound, you need to deduce the right number of positive ions and the right number of negative ions to get the total positive charge to equal the total negative charge.

Exam tip

Try this quick way of working out the formula of an ionic compound. Swap the numbers from the charges of the two ions to make the little numbers in the formula and then simplify if you can. For example, Al^{3+} and O^{2-} makes Al_2O_3. And Mg^{2+} and O^{2-} makes Mg_2O_2, which can then be simplified to MgO.

Example

Deduce the formula of ammonium sulfate.

- Ammonium ions have a charge of 1+, NH_4^+.
- Sulfate ions have a charge of 2–, SO_4^{2-}.
- You will need two ammonium ions to balance the 2– charge on the SO_4^{2-}.
- The formula is $(NH_4)_2SO_4$. Note how brackets are used when you have more than one of a complex ion.

Revision activity

To help you learn the formula and charge each of the important ions, make a set of cards. On the front, write the name of the ion, and on the back write the formula including its charge. Look at each ion's name and see if you can remember the formula, then turn over to see if you are correct.

Now test yourself

TESTED ☐

7 Deduce the formula of hydrogen sulfide.
8 Deduce the formula of methane, which is the simplest compound made from carbon and hydrogen.
9 Explain how you can use the Periodic Table to work out the charge on a calcium ion.
10 Deduce the formula of copper(II) carbonate.
11 Deduce the formula of zinc nitrate.

Answers on page 124

Particles in reactions – equations

Word equations and symbol equations

Chemical reactions can be described using word equations and symbol equations. The **reactants** are always shown on the left of the arrow and the **products** are always shown on the right of the arrow. For example, when methane burns in air it reacts with oxygen to produce carbon dioxide and water. The word equation and symbol equation for this reaction are:

methane + oxygen → carbon dioxide + water

$$CH_4 + O_2 \rightarrow CO_2 + H_2O$$

> **Reactants:** The chemicals that react together in a chemical reaction.
>
> **Products:** The chemicals that are produced in a chemical reaction.

Exam tip

When you are given a sentence or two about a reaction and asked to write a word or symbol equation, it is helpful to use different coloured highlighters to identify the names of the reactants and products. Alternatively, if you just have a pen you can underline the reactants and circle the products. This helps you leave out irrelevant words like **powder, liquid, solution,** etc.

For example:

When dilute hydrochloric acid reacts with potassium hydroxide solution, the products are potassium chloride and liquid water. Write a word equation for this reaction.

hydrochloric acid + potassium hydroxide → potassium chloride + water

Exam tip

Never mix words and symbols in an equation. You must also remember to use an arrow (→) in the middle of your chemical equations, not an equal sign (=). Equal signs should be saved for calculations.

State symbols

State symbols are sometimes added after the formulae in a symbol equation to show whether the reactants and products are solid, liquid, gas or dissolved in water (aqueous solution). Table 1.9 shows the state symbols you need to know.

> **State symbols:** Symbols used after the chemical formula of a substance to show whether it is a solid, liquid, gas or dissolved in water (aqueous solution).

Table 1.9 State symbols used in symbol equations

(s)	solid
(l)	liquid
(g)	gas
(aq)	aqueous solution (dissolved in water)

Example

Write a symbol equation to summarise the reaction when solid copper carbonate ($CuCO_3$) is added to a solution of hydrochloric acid (HCl). The products are copper chloride ($CuCl_2$) solution, water and carbon dioxide.

- $CuCO_3(s) + HCl(aq) \rightarrow CuCl_2(aq) + H_2O(l) + CO_2(g)$

Exam tip

When writing symbol equations, the examiner will probably expect you to know the formulae of some common substances (e.g. water and carbon dioxide) and their normal states of matter. If the examiner wants you to add state symbols, you will be asked to do so.

Balancing a symbol equation

It is essential for symbol equations to be balanced in order for them to make scientific sense. An unbalanced equation suggests that atoms have been made or destroyed during the reaction, which is impossible. Look at the following equation, which is unbalanced and therefore incorrect:

$$Cu + O_2 \rightarrow CuO \text{ (unbalanced)}$$

In this equation it appears as though one oxygen atom has been destroyed in this reaction. You need to count the number of atoms of each type on both sides of the equation and then work out the number to put in front of each chemical to balance the numbers of each type of atom.

$$2Cu + O_2 \rightarrow 2CuO \text{ (balanced)}$$

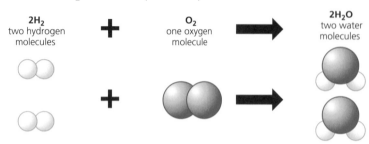

Figure 1.22 A balanced symbol equation and a particle diagram help to make sense of the way that atoms are involved in a chemical reaction

Now test yourself

TESTED

12 Write a word equation for the decomposition of calcium carbonate to form calcium oxide and carbon dioxide.

13 In a neutralisation reaction, sodium nitrate and water are produced when nitric acid solution reacts with sodium hydroxide. Represent this reaction using a word equation.

14 In a catalytic converter, NO reacts with CO to produce N_2 and CO_2. Write a balanced symbol equation for this reaction.

15 In the thermite reaction, aluminium powder reacts with iron(III) oxide powder (Fe_2O_3) to produce liquid iron and solid aluminium oxide (Al_2O_3). Write a balanced symbol equation, including state symbols.

Answers on page 124

Formulae and equations

Percentage composition

REVISED

The **percentage composition** of a compound can be calculated from the chemical formula, the A_r values of the elements and the M_r of the compound, using the following equation:

$$\text{percentage of element (E)} = \frac{(A_r \text{ of E}) \times (\text{number of atoms of E})}{M_r} \times 100$$

Percentage composition: The percentages by mass of each element in a compound. (N.B. The sum of the percentages must always equal 100.)

Example

Calculate the percentage of nitrogen in ammonium sulfate, $(NH_4)_2SO_4$.

● The M_r of ammonium sulfate is 132

● Percentage of nitrogen $= \dfrac{14 \times 2}{132} \times 100 = 21.2\%$

Calculating reacting masses in chemical equations

REVISED

Remember that the balancing numbers in an equation tell us the ratio of moles for each reactant and product. We can calculate the number of moles of a pure chemical (typically a solid) if we know its mass and M_r.

$$\text{Number of moles (mol)} = \frac{\text{mass of sample}}{M_r}$$

Exam practice answers and quick quizzes at **www.hoddereducation.co.uk/myrevisionnotesdownloads**

Calculate the mass of water produced when 8 g of methane burns:

$$CH_4 + O_2 \rightarrow CO_2 + H_2O$$

Table 1.10

Step 1: Balance the equation	$CH_4 + 2O_2 \rightarrow CO_2 + 2H_2O$
Step 2: Write the masses and M_r values under the relevant chemicals	$CH_4 + 2O_2 \rightarrow CO_2 + 2H_2O$ 8 g ? g $M_r = 16$ $M_r = 18$
Step 3: Calculate the number of moles of the 'known' chemical	CH_4 is the known chemical because we know its mass and its M_r. moles of $CH_4 = \frac{8}{16} = 0.5$ mol
Step 4: Use the 'balancing number ratio' to calculate the number of moles of the unknown chemical	From the equation we can see that 1 mole of CH_4 produces 2 moles of H_2O. Therefore, 0.5 mol of CH_4 will produce 1 mol of H_2O.
Step 5: Change the number of moles of the 'unknown' chemical to a mass	Mass of H_2O = mol of H_2O × M_r of H_2O = 1 × 18 = 18 g water produced

> **Exam tip**
>
> In step 2 of the example above, we only wrote the M_r values under the formulae of the chemicals that were mentioned in the question. Do not waste time calculating M_r values that you are not going to need in your calculation. (In this example, we did not need the M_r values for O_2 or CO_2.)

Volumes of gases in reactions

REVISED ☐

At room temperature (25 °C) and atmospheric pressure (1 atm), the volume of one mole of any gas is 24 dm³ (or 24 000 cm³). This number is called the **molar volume**. We can therefore link the number of moles of any gas to its volume using the following equation:

$$\text{moles of gas} = \frac{\text{volume (dm}^3)}{24}$$

> **Molar volume:** The volume that one mole of any gas occupies at a specified temperature and pressure.

> **Example**
>
> Calculate the number of moles of oxygen in 120 cm³.
> - Convert 120 cm³ to dm³ by dividing by 1000: 120 ÷ 1000 = 0.12 dm³
> - Moles of oxygen = $\frac{0.12}{24} = 0.005$ mol

> **Exam tip**
>
> Remember that 1 dm³ contains 1000 cm³. So if you are given a volume in cm³ in the exam, you will need to divide by 1000 to convert the volume into dm³ before using the equation.

The volume of a gas produced in a reaction can be measured using a gas syringe (Figure 1.23). It can be predicted using a moles calculation.

> **Example**
>
> Calculate the volume of gas produced when 1.2 g of magnesium reacts with hydrochloric acid: $Mg + HCl \rightarrow MgCl_2 + H_2$.
>
> | Step 1: Balance the equation | $Mg + 2HCl \rightarrow MgCl_2 + H_2$ |
> |---|---|
> | Step 2: Write the mass and A_r value under the relevant chemical | $Mg + 2HCl \rightarrow MgCl_2 + H_2$
1.2 g   ? dm³
$A_r = 24$ |

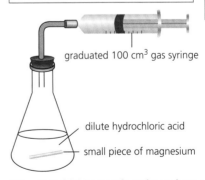

graduated 100 cm³ gas syringe

dilute hydrochloric acid

small piece of magnesium

Figure 1.23 Measuring the volume of a gaseous product during a reaction using a gas syringe

Step 3: Calculate the number of moles of the 'known' chemical	Mg is the known chemical because we know its mass and its A_r. Moles of Mg $= \dfrac{1.2}{24} = 0.05$ mol
Step 4: Use the 'balancing number ratio' to calculate the number of moles of the unknown chemical	From the equation we can see that 1 mole of Mg produces 1 mole of H_2. Therefore, 0.05 mol of Mg will produce 0.05 mol of H_2.
Step 5: Change the number of moles of the 'unknown' chemical to a volume	Volume of H_2 = mol of $H_2 \times 24$ $= 0.05 \times 24$ $= 1.2$ dm^3 H_2 produced

Percentage yield

The yield is the amount of the desired product that is made in a reaction. The maximum possible yield is what is calculated using a moles calculation. This is called the **theoretical yield** or **predicted yield**. In reality, the **actual yield** of the product that is made will always be less than the theoretical yield. This is because:

- The reactants might not be pure.
- Some of the product might be lost during the process, e.g. during filtration, spitting of a salt during evaporation/crystallisation.
- There may be unwanted side reactions.
- The reaction does not completely finish.

The percentage yield is calculated using the following formula:

$$\text{percentage yield} = \frac{\text{actual yield (g)}}{\text{predicted yield (g)}} \times 100$$

Theoretical yield (also called the **predicted yield**): The maximum possible mass of the desired product that could be made during a chemical reaction, if the reaction went perfectly.

Actual yield: The mass of the desired product that is actually obtained at the end of a chemical reaction following unavoidable losses.

Example

A student calculated that she should produce a theoretical yield of 6.0 g of $CuSO_4$ in a reaction. At the end of the process, she weighed her product and found that she had made 4.2 g. Calculate her percentage yield.

- Percentage yield $= \dfrac{4.2}{6.0} \times 100 = 70\%$

Exam tip

It does not matter what units are used for the actual and theoretical yields, as long as they are the same. The percentage calculation will still work, even if the masses are in kg or tonnes.

Now test yourself

16 Calculate the M_r for Fe_2O_3.
17 During the extraction of iron in the blast furnace, one of the reactions is $Fe_2O_3 + 3CO \rightarrow 2Fe + 3CO_2$. Calculate the mass of iron that can be obtained from 480 g of Fe_2O_3.
18 Calculate the volume of CO_2 (at room temperature and pressure) that would be obtained during this same reaction.
19 The extraction process was repeated using a different starting mass of iron oxide. The chemical engineer calculated that the theoretical yield of iron was 150 tonnes. The actual mass of iron obtained was 125 tonnes. Calculate the percentage yield in this reaction.

Answers on page 124

Typical mistake

Candidates often get the actual and theoretical yields the wrong way round in the equation. But remember that your answer must be less than 100%. If it is more than 100%, try swapping your numbers on the top and bottom of the fraction.

Chemical bonding

Ionic bonding

Ionic bonding occurs between a metal and a non-metal. The metal atom loses electrons to obtain a stable empty outer shell and becomes a positive **ion**. The non-metal atom gains electrons from a metal atom to obtain a stable full outer shell and becomes a negative ion. The positive metal ion and the negative non-metal ion are strongly attracted by **electrostatic forces** of attraction. These forces are also called ionic bonds. The ions form into a giant ionic **lattice structure** (see page 38).

Ionic bonding can be represented using dot-and-cross diagrams. The dots represent electrons from one atom and the crosses represent electrons from another atom.

> **Ionic bonding:** The kind of bonding that occurs between a metal atom and a non-metal atom. Electrons are transferred from the metal atom to the non-metal atom during ionic bonding.
>
> **Ion:** A particle with a positive or negative charge, which has been formed when an atom (or group of atoms) has gained or lost electrons.

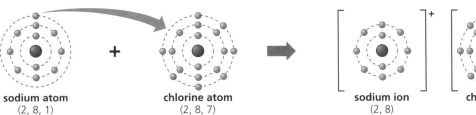

sodium atom (2, 8, 1)   chlorine atom (2, 8, 7)   sodium ion (2, 8)   chloride ion (2, 8, 8)

Figure 1.24 A dot-and-cross diagram showing the ionic bonding in sodium chloride (full electron configurations)

$$\text{Na}^\bullet + {}^\times_\times\text{Cl}^\times_\times \Rightarrow \left[\text{Na}\right]^+ \left[{}^\times_\times\text{Cl}^\times_\times\right]^-$$

(2, 8, 1)   (2, 8, 7)   (2, 8)   (2, 8, 8)

Figure 1.25 A dot-and-cross diagram showing the ionic bonding in sodium chloride (showing the outer electron shells only)

Remember that the group number of an atom tells you the number of electrons in its outer shell. When they react, metals lose all of their outer shell electrons, so this tells you their positive charge. For example, calcium is in group two, so it loses two electrons and forms an ion with a 2+ charge (Ca^{2+}). Non-metals gain enough electrons to fill their outer shell, so you can easily work out their negative charge. For example, oxygen is in group 6, so it must gain two electrons to fill its outer shell, so it has a charge of 2− (O^{2-}).

> **Electrostatic force:** An attractive force between two particles which have opposite charges, e.g. between a positive ion and a negative ion.
>
> **Lattice structure:** A 3D structure which consists of billions of atoms or ions in a regular arrangement.

magnesium fluoride

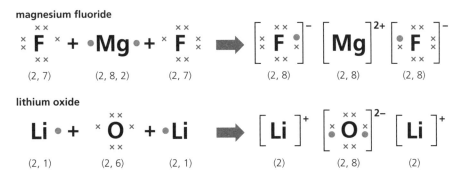

lithium oxide

Figure 1.26 Dot-and-cross diagrams (showing the outer shells only) for magnesium fluoride and lithium oxide

Covalent bonding occurs between two non-metal atoms. The two atoms share electrons in their outer shells so that they both achieve a stable full outer shell (rather than losing or gaining electrons from each other). The nuclei of both atoms have a strong electrostatic attraction to the shared pair of electrons, so covalent bonds are strong. Covalent bonding usually makes **molecules**, but sometimes covalently bonded substances have a **giant covalent structure**, e.g. diamond and graphite (see page 42).

Covalent bonding can be represented using dot-and-cross diagrams. Notice that the outer shells of the atoms overlap so that electrons can be shared. A **covalent bond** is a shared pair of electrons, normally represented with a dot and a cross next to each other. The electrons that are not shared are also arranged in pairs and are called **lone pairs**.

Covalent bonding: The kind of bonding that occurs between two non-metal atoms. Electrons are shared between the two atoms during covalent bonding.

Molecule: A cluster of non-metal atoms which are joined by covalent bonds.

Giant covalent structure: An extended structure of non-metal atoms that are joined by covalent bonds, e.g. silicon oxide, diamond.

Covalent bond: A pair of electrons that are shared between two non-metal atoms.

Lone pair: A pair of electrons that are not used in a covalent bond.

chlorine, Cl₂
Cl—Cl

hydrogen chloride, HCl
H—Cl

water, H₂O
H—O
|
H

ammonia, NH₃
H—N—H
|
H

methane, CH₄
H
|
H—C—H
|
H

Figure 1.27 Dot-and-cross diagrams, displayed formulae and molecular formulae for some covalent molecules. Notice the lone pairs in some molecules

Some molecules contain double or triple covalent bonds. Double bonds are formed when four electrons are found in a covalent bond. Triple bonds involve six electrons.

Exam tip

You should be aware that there are double bonds in the following molecules: O₂, CO₂, and all alkenes (see page 111). The only time you are likely to come across a triple bond is in N₂.

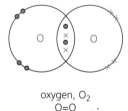
oxygen, O₂
O=O

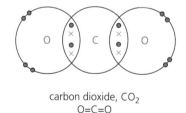
carbon dioxide, CO₂
O=C=O

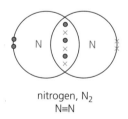
nitrogen, N₂
N≡N

Figure 1.28 Molecules that have double or triple covalent bonds

Now test yourself

TESTED ☐

20 Use the Periodic Table to work out the type of bonding in these substances: lithium iodide, sulfur dioxide, bromine, iron oxide.
21 Draw a dot-and-cross diagram for the ionic bonding in magnesium oxide.
22 Work out the formula of the ionic compound, sodium sulfide, using ideas about the charges on each ion.
23 Draw a dot-and-cross diagram for a molecule of hydrogen, H_2.

Answers on page 124

Answers on page 124

Revision activity

While you can usually figure out an unfamiliar dot-and-cross diagram in the exam, it is helpful to learn the more common ones. Write the name and formula on one side of a card and draw the dot-and-cross diagram on the back. Get someone to test your ability to recall the diagram from memory.

Which substances conduct electricity?

Substances that conduct as a solid

REVISED ☐

All metals conduct electricity, and all metals, except mercury, are solids at room temperature. Ionic compounds do not conduct electricity when they are solids. Covalent solids that are made of molecules (e.g. sulfur, iodine) do not conduct electricity, and neither do those that are made of giant structures (e.g. silicon dioxide), with the exception of graphite. Graphite is a form of pure carbon and it is the only non-metal element that conducts electricity.

Substances that conduct as a pure liquid

REVISED ☐

Mercury is a metal element that is a liquid at room temperature, but as it is a metal it still conducts electricity. Ionic substances conduct electricity if they are heated above their melting points. Covalent substances that are made of molecules do not conduct electricity (e.g. water, bromine, ethanol), as long as they are pure.

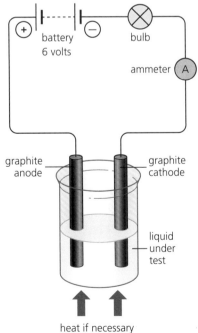

Figure 1.30 The apparatus that can be used to test if a substance conducts electricity as a liquid

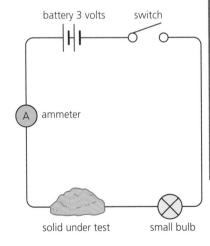

Figure 1.29 The apparatus that can be used to test if a solid conducts electricity

Substances that conduct when dissolved in water

Some ionic compounds dissolve in water to form a solution. For example, sodium chloride is soluble, but magnesium oxide is not. If an ionic compound dissolves in water, the solution produced will conduct electricity.

Solutions of covalent molecules do not conduct electricity, unless they react with the water to form ions. An example of this is when hydrogen chloride gas dissolves in water. In this example, the covalent HCl molecules split apart to produce two ions: H^+ and Cl^-. The ions in solution cause the solution to conduct electricity.

Electrolysis

When electricity is conducted through a **molten** ionic compound or a dissolved ionic compound, a chemical reaction occurs. The compound is **decomposed** (broken down) into elements, which are produced at each **electrode**. Breaking down a compound using electricity is called **electrolysis**. You will learn more about electrolysis on page 45.

Now test yourself

24 Group these pure solids into those which conduct electricity and those which do not: iron, iodine, calcium, cobalt, diamond, graphite.
25 Bromine is a liquid at room temperature. Does it conduct electricity? Explain your answer.
26 State and explain whether molten iodine will conduct electricity.
27 State and explain whether molten zinc chloride will conduct electricity.
28 State and explain whether an aqueous solution of iodine will conduct electricity.

Answers on page 124

Giant ionic structures

The structure of ionic compounds

Dot-and-cross diagrams are used to show the electron transfer from a metal atom to a non-metal atom during ionic bonding (see page 35). The positive metal ion and the negative non-metal ion formed in this process are strongly attracted by **electrostatic forces**, which are also known in this case as ionic bonds.

The positive and negative ions do not form small clusters like molecules of covalent substances. The ions are arranged into a giant **lattice structure**, which extends in three dimensions for billions of ions.

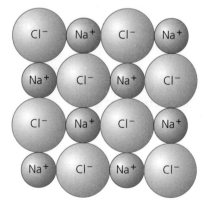 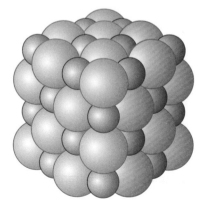

> **Molten:** A substance that has been melted from a solid into a liquid through heating.
>
> **Decomposition:** A kind of chemical reaction in which one reactant has been split into simpler substances. Energy is always needed to make decomposition reactions happen.
>
> **Electrodes:** The positive and negative parts of a circuit which can be connected to (or dipped into) a substance being investigated.
>
> **Electrolysis:** When a compound is broken down using electricity to produce elements.

> **Electrostatic force:** An attractive force between two particles which have opposite charges, e.g. between a positive ion and a negative ion.
>
> **Lattice structure:** A 3D structure which consists of billions of atoms or ions in a regular arrangement.

Figure 1.31 2D and 3D representations of the giant ionic lattice of sodium chloride

In Figure 1.31, you can see that it is easier to label the ions in a 2D ionic lattice diagram, but the 3D diagram gives a better representation of the actual structure and shows that each sodium ion is surrounded by six chloride ions.

Explaining the properties of ionic compounds

Ionic compounds have high melting points and high boiling points because a lot of energy is needed to break the strong electrostatic forces between the oppositely charged ions. Smaller ions, and ions that have a greater charge, have stronger attractions in the ionic lattice and therefore cause higher melting points.

Ionic compounds do not conduct electricity when they are solid because the ions are fixed in place and therefore cannot move to carry an electrical current.

Ionic compounds do conduct electricity when they are molten or dissolved because the ions are now free to move to carry an electrical current.

Now test yourself

29 Draw a labelled 2D diagram to show the structure of lithium bromide. You should include between six and 12 ions in your diagram.
30 Predict whether magnesium oxide will have a higher or lower melting point than sodium fluoride. Explain your answer.
31 Predict whether sodium fluoride will have a higher or lower melting point than potassium chloride. Explain your answer.
32 Explain why solid zinc chloride does not conduct electricity, but molten zinc chloride does conduct electricity.
33 Explain why pure water does not conduct electricity, but when a spatula of sodium chloride is added and the water is stirred, the liquid now conducts electricity.

Answers on pages 124–5

The structure of substances

Atoms, molecules and ions

All substances are made of three types of particles: **atoms**, **molecules** or **ions**:

- Atoms are neutral particles of an element; they are made from protons and neutrons in the nucleus, with electrons orbiting this nucleus in shells.
- Molecules are neutral particles of a non-metal element or covalent compound. Molecules are made from atoms which are bonded by covalent bonds (shared pairs of electrons).
- Ions are atoms (or groups of atoms) that have a positive or negative charge because they have gained or lost electrons.

> **Atom:** The smallest particle of an element, which has a central nucleus surrounded by orbiting electrons.
>
> **Molecule:** A cluster of non-metal atoms which are joined by strong covalent bonds.
>
> **Ion:** A charged particle, which is usually formed from an atom that has gained or lost electrons.

Bonding and structures

There are three types of bonding: ionic (page 35), covalent (page 36) and metallic (page 44). But there are four types of structures because covalent substances can either be made of molecules or **giant structures**. The four types of structures are summarised in Table 1.11.

> **Giant covalent structure:** An extended structure of non-metal atoms that are bonded by covalent bonds, rather than being made of molecules. Their structure will extend in three dimensions for billions of atoms.

Table 1.11 A summary of the bonding and properties in different types of substances

Bonding	Structure	Elements involved	Particles present	Electrons are...	Examples	Typical properties
Ionic	Giant ionic lattice	Metal and non-metal	Ions	Transferred from metal atom to non-metal atom	NaCl, MgO, CaCl$_2$	High melting point, do not conduct when solid, do conduct when molten or dissolved
Metallic	Giant metallic lattice	Metal elements and alloys	Positive metal ions and delocalised electrons	Delocalised and therefore free to move	Li, Fe, Zn, Cu, Al, steel	High melting point, conduct when solid, malleable
Covalent	Giant covalent structure	Non-metal	Atoms	Shared between atoms	Si, SiO$_2$, diamond, graphite	High melting point, do not conduct electricity (except graphite, which does conduct)
	Simple molecular covalent structure	Non-metal	Molecules	Shared between atoms	H$_2$O, CO$_2$, O$_2$, H$_2$	Low melting and boiling points, do not conduct electricity

Now test yourself

TESTED ☐

34 Substance X has a high melting point and it does not conduct when it is a solid, or when it is melted. Identify the type of bonding and structure that will be found in substance X.

35 Element Y forms a compound by reacting with oxygen. This compound has a high melting point and does not conduct when it is solid. When it is added to water, it dissolves and the solution conducts electricity. Identify the type of bonding and structure in the compound formed from Y and oxygen.

36 The oxide of element Z has a boiling point of −67 °C. Identify whether element Z is a metal or a non-metal. Explain your answer.

Answers on page 125

Revision activity

Copy out Table 1.11 but only fill in the first two columns. See if you can then fill in the remaining columns from memory.

Simple molecular substances

Types of covalent substances

REVISED ☐

Covalent bonding occurs when non-metal atoms share electrons. Covalent bonding occurs in non-metal elements (e.g. silicon, nitrogen, oxygen) and also in compounds made from two or more non-metal elements (e.g. water, carbon dioxide, silicon oxide).

Most covalent substances are made from **molecules**. These are called simple molecular compounds. A small number of covalent substances are made of **giant covalent structures**.

Covalent bonding: The kind of bonding that occurs between two non-metal atoms. Electrons are shared between the two atoms during covalent bonding.

Molecule: A cluster of non-metal atoms which are joined by covalent bonds.

Giant covalent structure: An extended structure of non-metal atoms that are joined by covalent bonds, e.g. silicon oxide, diamond.

Exam tip

It is easier to remember which substances are made of giant covalent structures and then you know that every other element or compound made from non-metal atoms is simple molecular. Learn these giant covalent substances: diamond, graphite, silicon, silicon oxide.

Properties of simple molecular covalent substances

Substances which are made of simple molecules do not conduct electricity. This is because simple molecular covalent substances contain no **delocalised electrons** or free moving ions which could move to carry an electrical current.

Most substances which are made of simple molecules have low melting and boiling points. This means that many of them are gases. For example, hydrogen, oxygen, nitrogen, carbon dioxide, methane, chlorine, sulfur dioxide, carbon monoxide. Some are liquids, such as water and ethanol. Some simple molecular substances are solids, but they usually melt at low temperatures, e.g. iodine and sulfur.

The reason for the low melting and boiling points is that the weak attractive forces between the molecules do not need much energy to be broken. These attractive forces are called **intermolecular forces of attraction**.

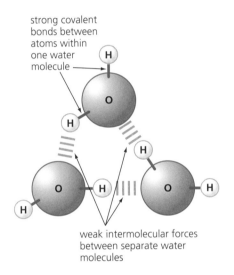

strong covalent bonds between atoms within one water molecule

weak intermolecular forces between separate water molecules

Figure 1.32 A diagram showing the strong covalent bonds between atoms in a molecule of water, and the weak intermolecular forces of attraction between the molecules (in blue)

The larger the molecule, the stronger the intermolecular forces of attraction. This means that the melting points of molecules with a larger relative formula mass (M_r) are usually higher than molecules with a lower M_r. This is shown in Table 1.12, where the compounds are listed in order of increasing M_r. You can see that the melting points increase with the M_r values, although the boiling point for CS_2 does not fit the normal pattern.

Delocalised electron: An electron which is not attached to an atom and is therefore able to move through a substance to carry an electrical current. Delocalised electrons are found in all metals and also in graphite.

Intermolecular force of attraction: A weak force that acts between two molecules.

Typical mistake

Candidates sometimes write about breaking chemical (covalent) bonds during melting and boiling. This is incorrect when describing simple molecular covalent substances. During melting and boiling, only the weak intermolecular forces of attraction are broken, not the strong covalent bonds between the atoms in the molecule, which stay intact.

Table 1.12 The M_r values, melting points and boiling points of some simple molecular compounds

Simple molecular substance	Formula	Relative molecular mass	Melting point in °C	Boiling point in °C
Carbon monoxide	CO	28	−205	−191
Hydrogen chloride	HCl	36.5	−114	−85
Carbon disulfide	CS_2	76	−112	46
Hydrogen bromide	HBr	81	−87	−67
Hydrogen iodide	HI	128	−51	−35

Now test yourself
TESTED ☐

37 What term is used to describe a cluster of non-metal atoms which are joined by sharing electrons?
38 Group the following substances into those which have a simple molecular covalent structure and those which have a giant covalent structure: bromine, hydrogen sulfide, silicon, graphite, carbon monoxide.
39 Which of the following will have the highest melting point: water, methane, diamond or iodine? Explain your answer.
40 Explain why butane (C_4H_{10}) is a gas at room temperature, but hexane (C_6H_{14}) is a liquid.

Answers on page 125

Diamond, graphite and fullerenes

Diamond and graphite are both made of pure carbon, but they are very different substances with very different properties and uses. This is because the carbon atoms are bonded to each other in different ways, which means diamond and graphite have different giant covalent structures. Different forms of a pure element in the same state are called **allotropes**. Diamond and graphite are allotropes of carbon.

> **Allotropes:** Different forms of a pure element in the same physical state.

Diamond

REVISED ☐

In diamond, each carbon atom is bonded to four other carbon atoms in a giant covalent structure. The properties of diamond and explanations for why it has these properties are summarised below.

- Diamond is very hard because the atoms are held in the giant covalent structure by many strong covalent bonds.
- Diamond has a very high melting point because lots of energy is needed to break the covalent bonds between the atoms.
- Diamond does not conduct electricity because there are no delocalised electrons or mobile ions that can move to carry a current.

Uses of diamond

Diamonds are used in cutting and drilling tools because they are so hard they can cut or drill through almost any substance. Their high melting point also prevents them from melting which might otherwise occur due to the heat caused by the friction created by drilling.

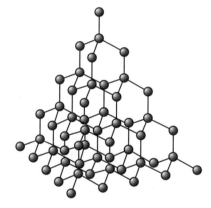

Figure 1.33 **The giant covalent structure of diamond**

Graphite

REVISED ☐

In graphite, each carbon atom is covalently bonded to three other carbon atoms in a giant covalent structure which is made from layers of hexagons. The fourth electron in the outer shell of each carbon atom is delocalised and can move through the structure. The properties of graphite and explanations for why it has these properties are summarised below.

- Graphite is very soft because the layers are only weakly attracted to each other and there are no covalent bonds between the layers. This means the layers can slide easily over each other.

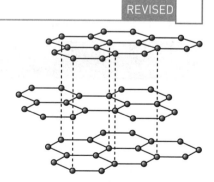

Figure 1.34 **The giant covalent structure of graphite**

- Graphite has a very high melting point because lots of energy is needed to break the covalent bonds between the atoms (this is the same as for diamond).
- Graphite is a good conductor of electricity (which is unusual for a non-metal) because it has delocalised electrons which can move through the structure to carry a current.

Uses of graphite

Graphite is used in pencils because it is so soft that layers of graphite will easily rub off onto the paper. Its softness also explains why graphite can be used as a solid lubricant to reduce the friction between surfaces, even if they get very hot. Graphite is also used to make electrodes for electrolysis because it conducts electricity and will not react with most other chemicals.

Fullerenes

REVISED

Fullerenes are a family of chemicals which are also allotropes of carbon. The first fullerene to be studied was C_{60}, which has a spherical structure made from pentagons and hexagons, just like a traditional football.

The discovery of C_{60} led to further research and discoveries. Carbon nanotubes are very long cylinders made from rolled-up sheets of hexagonally bonded carbon atoms. Graphene is a single sheet of hexagonally bonded carbon atoms.

Properties and uses of fullerenes

Fullerenes are currently being researched to explore their uses, for example in medicine to deliver drugs that kill cancer cells. As well as this, nanotubes might prove to be useful catalysts, and graphene is likely to be used in future electronic devices because it is an incredible conductor of electricity.

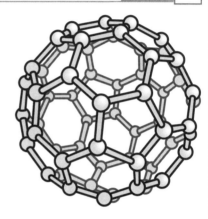

Figure 1.35 In C_{60}, each carbon atom is covalently bonded to three other atoms in a spherical structure.

Now test yourself

TESTED

41 Explain why graphite has a very high melting point.
42 Explain why diamond is used in cutting tools.
43 Why does graphite conduct electricity, but diamond does not?
44 Describe how the structures of graphite and C_{60} are similar and how they are different.
45 Compare the structures of carbon nanotubes and graphene.

Answers on page 125

The properties of metals

The metals are found on the left and in the middle of the Periodic Table (see page 15). Some metals are found low down in groups 3 and 4. Some metals have unusual properties, but the typical properties of metals are as follows:
- high melting points
- good conductors of electricity
- good conductors of heat
- high **densities**
- **malleable**.

Density: The mass of an object divided by its volume. For two objects that have the same size and shape, the heavier one will have the higher density.

Malleable: A substance is malleable if it can be beaten or hammered into a new shape. If an object is brittle (the opposite of malleable), hammering it will cause it to shatter.

Metals form positive ions when they react, whereas non-metals form negative ions (or molecules). Metal oxides are **bases**, which means that they react with acids to produce salts. If a metal oxide dissolves, it produces an **alkaline** solution. Non-metal oxides are **acidic** – if they dissolve, the solution has a pH of less than 7.

> **Base**: A substance that reacts with an acid, i.e. accepts a hydrogen ion (H⁺) in a reaction.

> **Alkali:** A base that dissolves in water. An alkaline solution has a pH greater than 7.
>
> **Acid:** A substance that reacts by giving away a hydrogen ion (H⁺) in a reaction. Acids produce salts when they react, and their solutions have a pH less than 7.

Bonding and structures of metals

REVISED

Metals have a giant structure which is made from a lattice structure of positive ions because the outer electrons from each metal atom are delocalised and free to move through the structure. The giant structure is held together by strong electrostatic forces between the positive metal ions and the delocalised electrons.

Explaining the properties of metals

REVISED

The key properties of metals are as follows:

- Metals have high melting points because lots of energy is needed to break the strong electrostatic attractions between the positive ions and the negative delocalised electrons.
- Metals are good conductors of electricity because the delocalised electrons are free to move towards the positive terminal in a circuit.
- Metals are good conductors of heat because the close-packed atoms and electrons transfer vibrations efficiently through the giant lattice.
- Metals have high densities because the atoms are closely packed together.
- Metals are malleable because the layers of atoms can slide over each other when a force is applied (Figure 1.37).

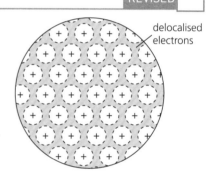

Figure 1.36 In a giant metallic lattice the positive ions are surrounded and glued together by delocalised electrons

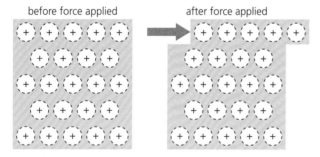

Figure 1.37 Metals are malleable because the layers of atoms can slide over each other

Alloys

REVISED

An **alloy** is a mixture of a metal element with another element. The other element it is usually another metal although sometimes it is the non-metal carbon. For example, steel is an alloy of iron and carbon.

Alloys have useful properties and are used in a variety of situations. One very useful property of many alloys is that they are often harder, stiffer and stronger than pure metals. This property can be explained by looking at their giant metallic structure.

> **Alloy:** A mixture of a metal element, with another element, which is usually a metal.

Exam practice answers and quick quizzes at **www.hoddereducation.co.uk/myrevisionnotesdownloads**

A pure metal is usually very malleable (see Figure 1.37). However, the presence of more than one type of atom in the lattice structure of an alloy means that the layers of atoms cannot slide over each other so easily, as the larger atoms stop the layers sliding. This is why they are usually stronger.

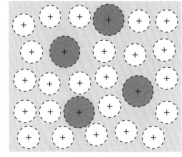

Now test yourself

TESTED

46 List three properties of a typical metal.
47 Describe how you could test to see if a solid substance is a metal.
48 Draw a 2D diagram of the structure of a metallic lattice.
49 Explain what makes metals good conductors of electricity.
50 A substance has a high melting point. It does not conduct electricity as a solid, but when it is melted it conducts electricity. Is it a metal? Explain your answer.

Answers on page 125

Figure 1.38 Alloys are less malleable than pure metals because the layers of atoms cannot slide over each other, owing to the presence of different sized atoms in the lattice

> **Revision activity**
>
> Make a summary poster for each of the four types of structures: metallic, ionic, giant covalent and simple molecular covalent.

Investigating electrolysis

The basics of electrolysis

REVISED

When electricity is used to split a compound into its elements, this is called **electrolysis**. The liquid that is being decomposed is called the **electrolyte** and this can either be a molten ionic compound or a solution of an ionic compound. Ions must be present in the electrolyte or it will not conduct electricity. The positive electrode is called the **anode** and this attracts **anions** (negatively-charged ions). The negative electrode is called the **cathode** and this attracts **cations** (positively-charged ions).

> **Electrolysis:** Splitting up a compound using electricity.
>
> **Electrolyte:** In electrolysis, the liquid that is being electrolysed.
>
> **Anode:** The positive terminal in a circuit.
>
> **Anion:** A negatively charged ion, e.g. Cl^- or OH^-.
>
> **Cathode:** The negative electrode in a circuit.
>
> **Cation:** A positively charged ion, e.g. H^+ or Cu^{2+}.

> **Exam tip**
>
> A good way to remember that cations are positive by thinking about pussytive cats (positive cations). Then remember that opposites attract, and the cations go to the cathode, so the cathode must be negative.

Electrode reactions

REVISED

When anions arrive at the anode, they lose electrons and form neutral atoms (which often pair up to form molecules of a gaseous element). This process is called **oxidation** and is represented in Figure 1.39, and in the following example **half equations**.

$$2Cl^- \rightarrow Cl_2 + 2e^-$$
$$2Br^- \rightarrow Br_2 + 2e^-$$

> **Oxidation:** Oxidation is the loss of electrons.
>
> **Half equation:** A chemical equation that involves electrons and shows either oxidation or reduction.
>
> **Reduction:** Reduction is the gain of electrons.

anode

Cl⁻ ion — attracted to cathode →

e⁻
electron given to anode

+

Cl atom

Figure 1.39 A chloride ion is attracted to the cathode and loses an electron to form a neutral atom

When the cations arrive at the cathode, they gain electrons and form neutral atoms (either metals or hydrogen gas). This process is called **reduction**, and is represented in Figure 1.40 and in the following example half equations:

$$Cu^{2+} + 2e^- \rightarrow Cu$$

$$2H^+ + 2e^- \rightarrow H_2$$

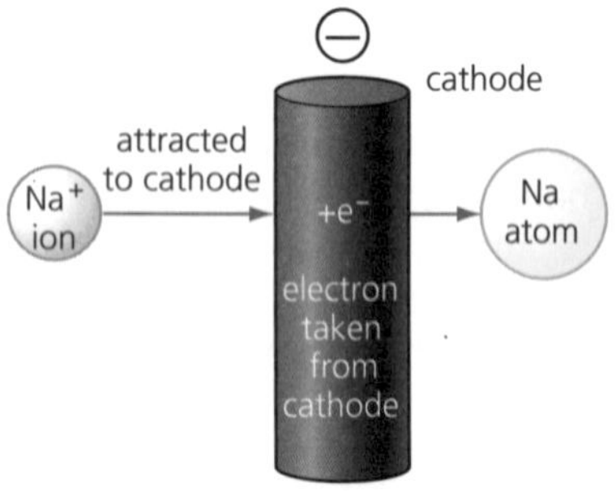

Figure 1.40 A sodium ion is attracted to the cathode and gains an electron to form a neutral atom

Electrolysis of molten ionic compounds

REVISED

In a molten ionic electrolyte there are only two types of ions, so predicting the products is easy. For example, zinc chloride ($ZnCl_2$) contains Zn^{2+} ions and Cl^- ions. The Zn^{2+} cations are reduced at the cathode to form Zn atoms. The Cl^- ions are oxidised at the anode to form Cl_2 molecules. The two half equations and overall equation are:

At the cathode: $Zn^{2+} + 2e^- \rightarrow Zn$

At the anode: $2Cl^- \rightarrow Cl_2 + 2e^-$

Overall: $ZnCl_2 \rightarrow Zn + Cl_2$

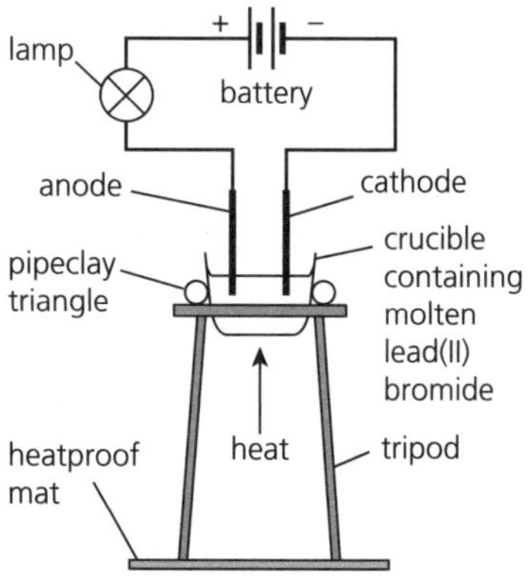

Example

Write half equations for each electrode in the electrolysis of molten lead bromide and an overall equation for the reaction.
- At the cathode: $Pb^{2+} + 2e^- \rightarrow Pb$
- At the anode: $2Br^- \rightarrow Br_2 + 2e^-$
- Overall: $PbBr_2 \rightarrow Pb + Br_2$

Figure 1.41 Electrolysing molten lead bromide

Electrolysing aqueous solutions

REVISED

Aqueous solutions of ionic compounds contain the anion and cation from the solute, but also contain some H^+ ions and OH^- ions from the water. The products of electrolysis depend on the reactivity of the metal and whether the anion is a **halide ion**.

Halide ion: A simple ion formed from a halogen atom (group 7) that has gained one electron, e.g. chloride (Cl^-), bromide (Br^-) or iodide (I^-).

Predicting the element produced at the cathode

Elements can be placed into a reactivity series by comparing their chemical reactions (see page 60). A simplified reactivity series of metals is shown on the right. Carbon and hydrogen are included even though they are non-metals.

potassium
sodium
lithium
calcium
magnesium
aluminium
carbon
zinc
iron
hydrogen
copper
silver
gold
platinum

Exam practice answers and quick quizzes at **www.hoddereducation.co.uk/myrevisionnotesdownloads**

If the metal cation from the ionic compound is lower than hydrogen in the series, the metal will be produced at the cathode. If the metal ion is more reactive than hydrogen, it will remain in the electrolyte and H^+ ions will be reduced to H_2 gas instead.

Predicting the element produced at the anode

If the anion from the ionic compound in the electrolyte is a halide ion, e.g. chloride (Cl^-) then the corresponding **halogen** will be produced. If the anion is not a halide (e.g. it could be the sulfate ion, SO_4^{2-}) then oxygen is produced from the hydroxide ions from the water:

$4OH^- \rightarrow 2H_2O + O_2 + 4e^-$ (oxidation)

Revision activity

Try to devise a mnemonic to help you to remember the reactivity series. It is often very useful in your exams, even though you will not be asked to recall it.

Halogen: An element from group 7 of the Periodic Table.

Required practical

Investigate the electrolysis of aqueous solutions

Method

1 The apparatus is set up as shown in Figure 1.42. The inverted test tubes should be full of water or the solution being tested.
2 The circuit is completed or the powerpack is turned on so that a current flows.
3 Observations are made at each electrode, including testing any gases produced.

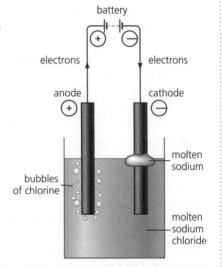

Figure 1.42 The apparatus used to electrolyse solutions when one or both products is a gas

Expected results

Aqueous solution (electrolyte)	Cathode	Anode	Ions left in electrolyte
Copper chloride, $CuCl_2(aq)$	Copper produced because it is less reactive than hydrogen $Cu^{2+} + 2e^- \rightarrow Cu$	Chlorine produced because the anion is a halide (Cl^-) $2Cl^- \rightarrow Cl_2 + 2e^-$	H^+ and OH^-
Sodium chloride, $NaCl(aq)$	Hydrogen produced because sodium is more reactive than hydrogen $2H^+ + 2e^- \rightarrow H_2$	Chlorine produced because the anion is a halide (Cl^-) $2Cl^- \rightarrow Cl_2 + 2e^-$	Na^+ and OH^- This leaves an alkaline solution of sodium hydroxide, NaOH
Sulfuric acid, $H_2SO_4(aq)$	Hydrogen produced because it is the only cation present $2H^+ + 2e^- \rightarrow H_2$	Oxygen produced because there is no halide ion present $4OH^- \rightarrow 2H_2O + O_2 + 4e^-$	H^+ and SO_4^{2-} Effectively, the water is electrolysed, leaving the H_2SO_4 behind
Copper(II) sulfate, $CuSO_4(aq)$	Copper produced because it is less reactive than hydrogen $Cu^{2+} + 2e^- \rightarrow Cu$	Oxygen produced because there is no halide ion present $4OH^- \rightarrow 2H_2O + O_2 + 4e^-$	H^+ and SO_4^{2-} This leaves an acidic solution of H_2SO_4

Now test yourself

51 Why must the electrolyte be a molten or dissolved ionic compound and not a simple molecular covalent compound?
52 What name is given to the positive electrode in electrolysis?
53 During electrolysis of molten lead bromide, what happens to the lead ion in order for metallic lead to be produced?
54 Predict the products of the electrolysis of an aqueous solution of potassium bromide.

Answers on page 125

Revision activity

Copy out the table in the required practical on the previous page but only fill in the first column. See if you can do the rest from memory.

Summary

- Relative formula mass (M_r) is calculated by adding up the relative atomic masses (A_r) of each atom in a formula.
- One mole of a substance contains 6×10^{23} formula particles, and has a mass in grams equal to the M_r of the substance.
- Word and symbol equations can be used to summarise reactions: reactants should always be on the left → products on the right.
- The percentage of an element in a compound can be calculated using the molecular formula, A_r values and M_r.
- The molar volume of any gas is $24\,dm^3$ ($24\,000\,cm^3$).
- The actual yield will always be less than the theoretical yield, and the percentage yield will always be less than 100%.
- Ionic bonding occurs between a metal atom and a non-metal atom. The metal atom loses electrons and becomes a positive ion. The non-metal atom gains electrons and becomes a negative ion. The oppositely-charged ions attract in a giant ionic lattice.
- Ionic compounds have high melting points and conduct electricity only when molten or dissolved.

- Covalent bonding occurs between two non-metal atoms. They share electrons in their outer shells to achieve full outer shells.
- Covalent bonding usually produces simple molecular covalent substances which have low boiling points and do not conduct electricity.
- Some covalent substances have a giant covalent structure, so they have high melting points.
- Diamond, graphite and fullerenes are all allotropes of carbon but they have different properties and uses because their structures are different.
- Metallic bonding involves the metal atoms losing their outer electrons, which become delocalised and glue the positive metal ions together in a giant lattice.
- Alloys are mixtures of a metal element with another element and they have useful properties.
- Electrolysis of a molten compound produces the elements that the compound is made from.
- Electrolysis of an aqueous ionic compound produces either the metal or hydrogen at the cathode. At the anode, oxygen or a halogen is produced.

Exam practice

1 Magnesium is a metal that can be burned in air. When it burns, it reacts with oxygen to produce magnesium oxide, MgO.
  a Write a word equation for the combustion of magnesium. [1]
  b Write a balanced symbol equation for the combustion of magnesium. [2]
  c Calculate the percentage by mass of magnesium in magnesium oxide, MgO. [2]
  d Draw a dot-and-cross diagram to show the ionic bonding in magnesium oxide. [2]
  e Explain why magnesium oxide has a high melting point and conducts electricity when molten but not when it is a solid. You should refer to the structure of magnesium oxide in your answer. [3]

2 Copper can be extracted from copper oxide by heating with carbon. A student heated 15.9 g of copper oxide with carbon.

   a Copy and complete the equation for the reaction of copper oxide with carbon. State symbols are not required.

      ___ CuO + ___ C → ___ Cu + ___ CO_2 [1]

   b Calculate the number of moles of copper oxide in 15.9 g.   [1]

   c Deduce the number of moles of copper produced in this reaction from 15.9 g of copper oxide.   [1]

   d Calculate the mass of copper produced.   [1]

3 Ethane is a hydrocarbon compound that has the formula C_2H_6. It can be used as a fuel.

   a Draw a dot-and-cross diagram to show the bonding in a molecule of ethane.   [2]

   b Ethane is a gas with a low boiling point. Explain why ethane has a low boiling point.   [2]

   c Pentane is also a hydrocarbon compound, and it has the molecular formula C_5H_{12}. Pentane is a liquid at room temperature. Explain why pentane is a liquid but ethane is a gas.   [2]

   d In an experiment, 0.1 moles of pentane was completely burned to produce carbon dioxide and water, according to the following equation.

      $C_5H_{12} + 8O_2 \rightarrow 5CO_2 + 6H_2O$

      Calculate the volume of carbon dioxide gas that is produced when 0.1 moles of pentane is completed burned at room temperature and pressure.   [3]

4 This question is about carbon and its compounds. Carbon is found as several different allotropes, including diamond, graphite and C_{60} fullerene.

   a Identify the three allotropes of carbon shown in the diagrams below.

      [3]

   b Explain why diamond and graphite have high melting points, but carbon dioxide has very low melting and boiling points.   [6]

   c Compare the hardness of diamond and graphite. You should refer to their bonding and structures in your answer.   [6]

5 Ionic substances can be electrolysed to obtain chemical elements.

   a Copy and complete the following sentences about electrolysis. Use the words from the list once, more than once, or not at all.   [5]

      anode   **electrolyte**   **solid**   **melted**   **dissolved**   **non-metal**   **metal**

      Ionic compounds are made when a non-metal element bonds with a _____ element. Ionic compounds conduct electricity if they are _____ or _____, but not if they are _____. During electrolysis, the liquid that is being split into elements is called the _____.

   b i Molten lead bromide is electrolysed. Identify the products that will be formed at the anode and cathode.   [2]

      ii Write a half equation for the formation of lead during the electrolysis of molten lead bromide, starting with the lead(II) ion, Pb^{2+}.   [2]

      iii Identify whether the lead ion has been oxidised or reduced in this process and explain your answer.   [2]

   c Copy and complete the table to show the products of electrolysis of the following aqueous solutions.   [4]

Aqueous compound	Product at anode	Product at cathode
Copper sulfate		
Sodium chloride		

Answers and quick quizzes online

ONLINE ☐

2 Inorganic chemistry

Properties of group 1 metals

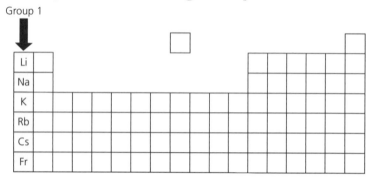

Figure 2.1 Group 1 is on the left of the Periodic Table

Exam tip

You only need to remember the chemical equations for lithium because the other alkali metals react in the same way. Just substitute the word *lithium* or symbol *Li* with the name of any other element in group 1.

Reactions with water

REVISED

The elements in group 1 are called the **alkali metals**. This is because they all react with water to produce an alkaline solution, plus hydrogen gas. Table 2.1 summarises the reactions.

Alkali metal: An element in group 1 of the Periodic Table.

Table 2.1 The reactions of the alkali metals with water

Element	Reaction with water	Word and symbol equations
Lithium, Li	• Floats, fizzes, moves around slowly • A flammable gas (H_2) is produced • An alkaline solution is produced (lithium hydroxide, LiOH)	Lithium + water → lithium hydroxide + hydrogen $2Li + 2H_2O → 2LiOH + H_2$
Sodium, Na	• Floats, fizzes more rapidly, melts into a ball, moves around rapidly • A flammable gas (H_2) is produced • An alkaline solution is produced (sodium hydroxide, NaOH)	Sodium + water → sodium hydroxide + hydrogen $2Na + 2H_2O → 2NaOH + H_2$
Potassium, K	• Floats, pops and sparks • A flammable gas (H_2) is produced • An alkaline solution is produced (potassium hydroxide, KOH)	Potassium + water → potassium hydroxide + hydrogen $2K + 2H_2O → 2KOH + H_2$

Trend in reactivity

REVISED

The observations in Table 2.1 show that the alkali metals get more reactive as you go down the group. This is also seen in their reactions with oxygen in the air. For example, when freshly cut with a sharp knife, lithium reacts with oxygen in under a minute to form a layer of lithium oxide. Sodium similarly oxidises within 20 seconds, and potassium oxidises almost immediately.

The group 1 elements react in a similar way because they all have one electron in their outer shell and they lose this outer electron when they react with other substances to achieve a stable empty outer shell (see pages 16–17). How easily they lose this outer electron determines their reactivity. The lower down in group 1 an element is, the larger the metal atom. This means that the outer electron is further from the positive nucleus and therefore it is held less strongly and lost more easily, so the metal is more reactive.

Because we know the similarities between alkali metals and how their properties affect their behaviour, we can make predictions about other alkali metals before testing them. For example, we can predict the reactions of rubidium and caesium with water.

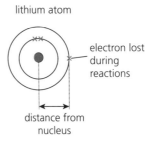

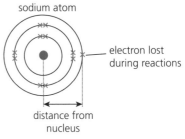

Figure 2.2 The electron configurations of a lithium atom and a sodium atom

Example

Use your knowledge of the reactions of Li, Na and K with water to predict the observations when Rb reacts with water.
- Write a symbol equation for the reaction:

$2Rb + 2H_2O \rightarrow 2RbOH + H_2$

- Rb will react more vigorously with water than K, so it will explode.

Trends in physical properties

REVISED

As you go down group 1, there are also clear trends in the **physical properties**. This means we can also make predictions about the physical properties of other elements in the group. Table 2.2 shows the melting points of three alkali metals.

Table 2.2 The melting points of lithium, sodium and potassium

Element	Relative atomic mass (A_r)	Melting point (°C)
Lithium, Li	7	180
Sodium, Na	23	98
Potassium, K	39	64

Physical properties: Descriptions of the ways that a substance characteristically behaves when not reacting with other chemicals. For example, its melting point, electrical conductivity, whether it is brittle, strong, hard, etc.

Example

Predict the melting point of rubidium.
- Rubidium is below potassium in group 1, so its melting point will be lower than 64 °C.
- Looking at the trend in the table, the difference between the elements decreases each time. Between 180 and 98 the difference is −82; between 98 and 64 the difference is −34.
- This suggests a decrease of 20 °C from 64 °C, giving 44 °C. (The actual melting point of Rb is 39 °C.)

Now test yourself

TESTED

1 Identify the chemical product of the reaction between rubidium and water that makes the solution alkaline.
2 Why do the elements in group 1 react in similar ways with water?
3 Explain why the reaction between potassium and oxygen is faster than the reaction between sodium and oxygen.
4 Suggest why sodium and potassium turn into molten metal when they react with water, whereas lithium does not.

Answers on page 125

Properties of group 7 elements

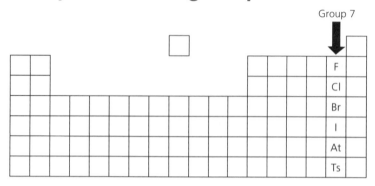

Figure 2.3 Group 7 is near the right of the Periodic Table

> **Halogen:** An element that is found in group 7 of the Periodic Table.
>
> **Halide ion:** A single negative ion that is formed from an element in group 7 of the Periodic Table, e.g. Br^-.
>
> **Diatomic:** A molecule which is made from two atoms that are joined by a covalent bond.

Physical properties

The elements in group 7 are called the **halogens**. They all have seven electrons in their outer shell, so they gain or share one electron when they react to achieve a stable full outer shell. When a halogen atom gains one electron, it forms a **halide ion**, for example Cl^-. When a halogen atom bonds with another non-metal atom it forms one covalent bond. This explains why the group 7 elements exist as **diatomic** molecules, for example Br_2. For more detail on covalent bonding, see page 36.

Table 2.3 The physical properties of three of the halogens

Element	A_r	Appearance at room temp.	Colour of vapour	Melting point (°C)	Boiling point (°C)
Chlorine, Cl_2	35.5	Pale green gas	Green	−101	−35
Bromine, Br_2	80	Brown liquid	Orange	−7	58
Iodine, I_2	127	Dark grey solid	Purple	114	183

The larger the relative formula mass (M_r) of the halogen molecule, the stronger its intermolecular forces of attraction (see page 41). This increase causes higher melting and boiling points.

Much like the elements in group 1, as the physical properties of these elements are similar and there is a trend, you can predict the appearance and properties of other halogens.

> **Revision activity**
>
> Cover up the columns for appearance and colour of vapour, and test to see if you can remember them.

> **Example**
>
> Predict the appearance and boiling point of fluorine, which is above chlorine in group 7.
> - Fluorine will be a gas because the boiling points decrease as you go up group 7 and chlorine, which sits below fluorine, is already a gas.
> - It is likely to be very pale or colourless because the elements get paler as you go up group 7.
> - A sensible prediction for its boiling point is around −100 °C. This is because as you go from I_2 to Br_2 the boiling point decreases by 125. From Br_2 to Cl_2 the decrease is 93. A sensible decrease from Cl_2 to F_2 would be around 60, so an answer around −90 to −100 is justified.
>
> (The actual boiling point of F_2 is −188 °C.)

5 What is meant by the term 'diatomic'?
6 Astatine has never been isolated as an element because it is radioactive and very rare. However, use your knowledge of other group 7 elements to predict its appearance.
7 Draw a dot-and-cross diagram of a molecule of chlorine.
8 Explain why chlorine is a gas at room temperature but bromine is a liquid.

Answers on pages 125–6

Reactions of the halogens

Reactions with hydrogen

REVISED

The halogens all react with hydrogen and these reactions can be used to demonstrate the reactivity of the halogens. An example of one of these reactions is:

hydrogen + chlorine → hydrogen chloride

$$H_2 \quad + \quad Cl_2 \quad \rightarrow \quad 2HCl$$

Using hydrogen reactions to observe reactivity:

- The first halogen in the group, fluorine, reacts instantly and explosively with hydrogen to form hydrogen fluoride.
- Chlorine can be mixed with hydrogen in the dark, but as soon as the mixture is exposed to bright light, it explodes to form hydrogen chloride.
- Bromine will usually only react with hydrogen if a burning splint is added to the mixture.

These observations prove that the halogens get less reactive as you go down the group.

Displacement reactions

REVISED

When a solution of the element chlorine is added to a solution of the compound potassium bromide, a reaction occurs and bromine is produced, along with potassium chloride. The chlorine has displaced the bromine because it is more reactive. This is a **displacement reaction**.

chlorine + potassium bromide → bromine + potassium chloride

$$Cl_2 \quad + \quad 2KBr \quad \rightarrow \quad Br_2 \quad + \quad 2KCl$$

Displacement reaction: A reaction where a more reactive element takes the place of a less reactive element in a compound.

Studying the displacement reactions of the halogens provides evidence about the trend in reactivity, as summarised in Table 2.4. There is no point in adding chlorine solution to sodium chloride, hence the grey cells in the table.

Table 2.4 Observations and equations for the displacement reactions of the halogens

	Sodium chloride	Sodium bromide	Sodium iodide
Chlorine solution (pale green)		An orange solution of bromine is produced $Cl_2 + 2NaBr \rightarrow Br_2 + 2NaCl$ $Cl_2 + 2Br^- \rightarrow Br_2 + 2Cl^-$	A brown solution of iodine is produced $Cl_2 + 2NaI \rightarrow I_2 + 2NaCl$ $Cl_2 + 2I^- \rightarrow I_2 + 2Cl^-$
Bromine solution (orange)	No reaction (bromine is not reactive enough to displace chlorine)		A brown solution of iodine is produced $Br_2 + 2NaI \rightarrow I_2 + 2NaBr$ $Br_2 + 2I^- \rightarrow I_2 + 2Br^-$
Iodine solution (dark brown)	No reaction (iodine is not reactive enough to displace chlorine)	No reaction (iodine is not reactive enough to displace bromine)	

Revision activity

Copy out Table 2.4 with the headings but leave all the other nine cells blank. Using your understanding of reactivity and displacement reactions, see if you can fill in the cells.

It does not matter whether the halide ion comes from a solution of sodium chloride or potassium chloride. The group 1 metal ion is a **spectator ion**, so we can ignore it in the equation, leaving us with an **ionic equation**.

Spectator ion: An ion that does not take part in a reaction, so it can be omitted from the ionic equation for that reaction.

Ionic equation: A symbol equation that includes one or more ions.

Explaining the trend in reactivity of the halogens

REVISED

The smaller the halogen atom, the more reactive it is. This is because halogen atoms react by gaining an electron to fill the outer shell. The smaller the atom, the stronger the attraction will be to the electron which is gained, because the extra electron will be joining a shell closer to the nucleus. This close proximity makes it easier to gain the electron.

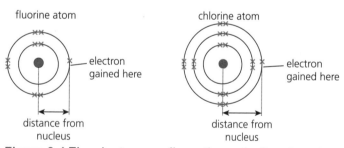

Figure 2.4 The electron configurations of a fluorine atom and a chlorine atom

Oxidation and reduction in displacement reactions

REVISED

When a halide ion loses electrons to become a halogen atom (in a molecule), it has been oxidised. When a halogen atom (in a molecule) gains an electron to form a halide ion, it has been reduced. This can be remembered by the phrase: **O**xidation **I**s **L**oss of electrons; **R**eduction **I**s **G**ain of electrons, giving you the mnemonic **OIL RIG**.

Because reduction and oxidation are occurring at the same time, this is called a **redox reaction**.

Redox reaction: A reaction in which electrons are transferred from one atom or ion to another atom or ion. This means something is **reduced** and something else is **oxidised**.

Exam practice answers and quick quizzes at **www.hoddereducation.co.uk/myrevisionnotesdownloads**

Now test yourself

9 Predict whether fluorine solution will react with a solution of potassium chloride. Explain your answer.
10 Write a word equation for the reaction of bromine solution with a solution of potassium iodide.
11 Write a balanced symbol equation for the reaction of chlorine solution with a solution of lithium bromide.
12 Write an ionic equation for the reaction of chlorine solution with a solution of potassium iodide.
13 Use what you know about electrons to explain why bromine atoms cannot oxidise chloride ions.

Answers on pages 125–6

Oxidation and reduction

Gaining or losing oxygen

As we have just seen, **oxidation** and **reduction** always occur together. The combined process is therefore called redox. Oxidation occurs when a substance has gained oxygen during a reaction. Reduction occurs when a substance has lost oxygen during a reaction. Redox can therefore be defined as being when an oxygen atom is transferred.

Combustion reactions are obvious oxidation reactions, for example:

methane + oxygen → carbon dioxide + water

$$CH_4 + 2O_2 \rightarrow CO_2 + 2H_2O$$

Respiration and rusting are also oxidation reactions, because O_2 is on the left of the equations, showing that oxygen atoms are being gained by another substance in the reactions:

$$C_6H_{12}O_6 + 6O_2 \rightarrow 6CO_2 + 6H_2O$$

$$4Fe + 3O_2 \rightarrow 2Fe_2O_3$$

Extracting metals from their oxide compounds involves reduction, for example:

iron(III) oxide + carbon monoxide → iron + carbon dioxide

$$Fe_2O_3 + 3CO \rightarrow 2Fe + 3CO_2$$

In this reaction, the iron in iron(III) oxide has been reduced by the carbon monoxide. We say that the carbon monoxide is the **reducing agent**.

In the thermite reaction, iron(III) oxide reacts with aluminium:

iron(III) oxide + aluminium → iron + aluminium oxide

$$Fe_2O_3 + 2Al \rightarrow 2Fe + Al_2O_3$$

In this reaction, the iron in iron(III) oxide has been reduced to iron by the aluminium. But we can also see that the aluminium has been oxidised (it has gained oxygen) by the iron(III) oxide. The iron(III) oxide is therefore the **oxidising agent**.

> **Oxidation:** A process in which a substance gains oxygen (or loses electrons).
>
> **Reduction:** A process in which a substance loses oxygen (or gains electrons).

> **Reducing agent:** A chemical which reduces another chemical by removing oxygen from it (or giving it electrons).
>
> **Oxidising agent:** A chemical which oxidises another chemical by giving it oxygen (or removing electrons).

Gaining or losing electrons

You also need to remember that oxidation and reduction can be defined in terms of electron transfer. This is sometimes more relevant than looking for oxygen atoms in a reaction. **O**xidation **I**s **L**oss of electrons, and **R**eduction **I**s **G**ain of electrons. Remember the mnemonic **OIL RIG**.

Half equations are used to show either reduction (if the electrons are on the left of the arrow) or oxidation (if the electrons are on the right).

> **Example**
>
> Write a half equation for the reduction of chlorine molecules to Cl^- ions.
> ● $Cl_2 + 2e^- \rightarrow 2Cl^-$

14 Identify what has been reduced in this reaction:
$$2CuO + C \rightarrow 2Cu + CO_2$$
15 Identify the reducing agent in the same reaction.
16 Identify whether reduction or oxidation is occurring in this half equation:
$$Fe^{3+} + 3e^- \rightarrow Fe$$
17 Write a balanced half equation for the oxidation of Br^- ions to Br_2 molecules.

Answers on page 126

The air

The composition of dry air

REVISED ☐

The four most abundant gases in dry air are summarised in Table 2.5.

Measuring the percentage of oxygen in air

Several methods can be used to measure the volume of oxygen in air, but they all rely on reacting a substance with oxygen to form a solid oxide compound and then measuring the decrease in the volume of the gas left behind. When the oxygen atoms combine with another element to form a solid oxide, they effectively take up no space.

Table 2.5 The most abundant gases in dry air

Gas	Approximate percentage of dry air
Nitrogen	78
Oxygen	21
Argon	0.9
Carbon dioxide	0.04

Required practical

Using the oxidation of copper to determine the approximate percentage by volume of oxygen in air

Method

1 The apparatus was set up as shown in Figure 2.5. Clamps were used to support the two gas syringes. One of the gas syringes contained $100\,cm^3$ of air; the other contained no air.

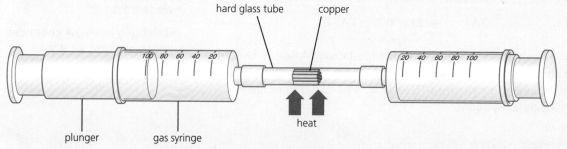

Figure 2.5 The apparatus used to determine the amount of oxygen in air

2 A Bunsen burner was used to heat the copper. The plunger of the left syringe was pushed in to make the air pass over the heated copper and into the right syringe. Oxygen atoms from the air reacted with the copper to make solid copper oxide.

3 The plunger of the right syringe was then pushed in to pass the air back across the copper.

4 Steps 2 and 3 were repeated until there was no decrease in the volume of gas remaining.

5 The apparatus was allowed to cool.

Results

After cooling, the volume of gas remaining was measured at 79 cm³, which was the total of all the other gases except oxygen. This gave: 100 − 79 = 21% oxygen.

Using iron wool to measure the percentage by volume of oxygen in air

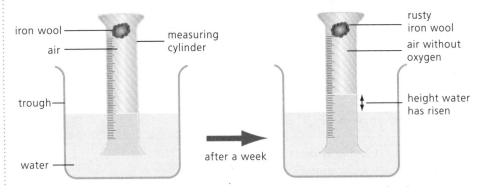

Figure 2.6 Using iron wool to determine the amount of oxygen in air

In this experiment, over the course of a few days, the iron reacts with oxygen from the air trapped in the measuring cylinder. This reduces the volume of the air trapped in the measuring cylinder by 21%.

Using phosphorus to measure the percentage by volume of oxygen in air

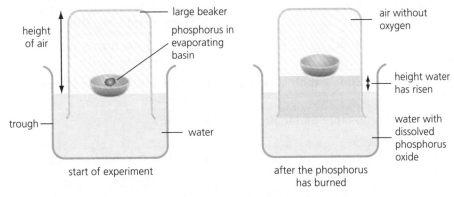

Figure 2.7 Using phosphorus to determine the amount of oxygen in air

White phosphorus is a very reactive non-metal, and it spontaneously ignites and burns in dry air. When it does this, it uses up the oxygen in the air trapped inside the beaker, causing the water level to rise. In this experiment, the volume of air trapped also decreases by 21%.

Now test yourself

TESTED

18 Name the four most abundant gases in dry air.
19 State the percentage of oxygen in dry air.
20 Explain why the volume of air decreases when it reacts with magnesium in a sealed system.
21 Describe how iron wool can be used to measure the percentage of oxygen in air.

Answers on page 126

Oxygen

Testing for oxygen

To test for oxygen, insert a glowing splint into a test tube containing the gas. The positive result for the presence of oxygen is that the splint relights.

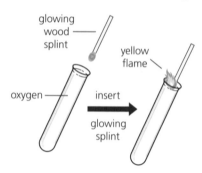

glowing wood splint

yellow flame

oxygen

insert

glowing splint

Figure 2.8 Testing a gas to see if it is oxygen

Reactions of oxygen with different elements

Oxygen will react with many different metal and non-metal elements, depending on the conditions. These reactions are summarised in Table 2.6. Metals are shown in blue and non-metals are in red.

Table 2.6 A summary of the reactions of oxygen with different elements

Element	Product	Type of bonding and structure in the oxide	pH when dissolved in water
Magnesium	Magnesium oxide, MgO (a white powder)	Ionic	pH 8; forms magnesium hydroxide, $Mg(OH)_2$
Sodium	Sodium oxide, Na_2O (a white powder)	Ionic	pH 11–14; forms sodium hydroxide, NaOH
Carbon	Carbon dioxide, CO_2 (a colourless gas)	Simple molecular covalent	pH 5; produces carbonic acid, H_2CO_3
Hydrogen	Water, H_2O (a colourless liquid)	Simple molecular covalent	pH 7
Sulfur	Sulfur dioxide, SO_2 (a colourless gas)	Simple molecular covalent	pH 2–3; produces sulfuric(IV) acid, H_2SO_3

You can see that non-metals react with oxygen to form covalent oxides, which will dissolve in water to make **acids** (except for water and silicon oxide). Metals react with oxygen to form ionic oxides, which are **bases**. If a base dissolves in water, it forms an **alkali**.

Acid: A substance that reacts by giving away a hydrogen ion (H^+) in a reaction. Acids produce salts when they react, and their solutions have a pH less than 7.

Base: A substance that reacts with an acid, accepting a hydrogen ion (H^+) in a reaction.

Alkali: A base that dissolves in water. An alkaline solution has a pH greater than 7.

Now test yourself

22 Describe the test and positive result for the presence of oxygen.
23 Write a word equation for the reaction between sodium and oxygen.
24 Write a balanced symbol equation for the reaction between magnesium and oxygen.
25 Element X is heated with oxygen. The oxide produced is a solid with a high melting point. When the oxide is dissolved in water, the solution has a pH of 10. Deduce whether element X was a metal or a non-metal and explain your answer.

Answers on page 126

Carbon dioxide

Properties and uses of carbon dioxide

Carbon dioxide is made of small molecules which have weak intermolecular forces of attraction. This weak attraction means it has a very low boiling point. The gas can only be made to condense into a liquid under high pressure, for example in fire extinguishers. Carbon dioxide is used to put out fires because it is denser than air and so smothers the fire and prevents oxygen from reaching the flames.

If the gas is cooled below −78°C it turns into solid carbon dioxide, which is known as 'dry ice'. If dry ice is allowed to warm up above −78°C, it **sublimes** into carbon dioxide gas. Dry ice is used as a refrigerant to keep things cold.

> **Sublime:** When a solid turns directly into a gas without melting into a liquid first.

Carbon dioxide is soluble in water and some of the dissolved carbon dioxide molecules react with the water to form carbonic acid, H_2CO_3. This is why fizzy drinks often have a sharp taste.

Making carbon dioxide

Carbon dioxide can be made through some simple chemical reactions.

The reactions of metal carbonates with acids

The general word equation for these reactions is:

acid + carbonate → salt + water + carbon dioxide

For example:

hydrochloric acid + calcium carbonate → calcium chloride + water + carbon dioxide

$$2HCl + CaCO_3 \rightarrow CaCl_2 + H_2O + CO_2$$

The thermal decomposition of metal carbonates

When heated, many carbonate compounds **thermally decompose** to form an oxide, which releases carbon dioxide. The general equation for this reaction is:

metal carbonate → metal oxide + carbon dioxide

> **Thermal decomposition:** When a compound breaks down into simpler substances once heated.

For example:

copper(II) carbonate → copper oxide + carbon dioxide

$$CuCO_3 \rightarrow CuO + CO_2$$

Testing for carbon dioxide

To test for carbon dioxide, you should bubble it through limewater, which is a dilute solution of calcium hydroxide. The positive result for the presence of carbon dioxide is the limewater turning cloudy/milky.

> **Typical mistake**
>
> When asked for the test for carbon dioxide, students will often write that it puts out a burning splint. However, this is not a positive test for carbon dioxide, because many other gases will also put out a burning splint: nitrogen, neon, argon, etc.

Carbon dioxide as a pollutant

Burning fossil fuels releases carbon dioxide into the atmosphere. Carbon dioxide is a **greenhouse gas**, which means that it contributes to the **greenhouse effect**, absorbing radiation from the Sun and Earth in the atmosphere. The vast majority of scientists agree that the increasing level of carbon dioxide in the atmosphere is causing **global warming** (climate change).

> **Typical mistake**
>
> Students often confuse the greenhouse effect with global warming. The greenhouse effect is natural; it maintains a warm atmosphere on Earth and is essential to almost all living things. However, additional global warming caused by human activities is likely to have devastating consequences.

Now test yourself

TESTED

26 Write a word equation for the reaction of sulfuric acid with sodium carbonate.
27 Write a symbol equation for the thermal decomposition of calcium carbonate ($CaCO_3$) to produce calcium oxide (CaO) and carbon dioxide.
28 Write a balanced symbol equation for the reaction of hydrochloric acid (HCl) with copper(II) carbonate ($CuCO_3$).
29 What is meant by the term 'greenhouse gas'?

Answers on page 126

> **Greenhouse gas:** A gas that absorbs infrared radiation in the atmosphere, contributing to the greenhouse effect.
>
> **Greenhouse effect:** A natural process in the atmosphere where certain gases absorb radiation from the surface of the Earth, keeping the atmosphere at a warm and stable temperature.
>
> **Global warming:** The steady rise in global atmospheric temperatures over long periods of time (e.g. decades).

The reactions of metals

Metals and acids

Observing the reactions of different metals can be done to place the metals into a list of decreasing reactivity. This list is described as a **reactivity series**. Metals that are higher than hydrogen in the reactivity series will react with dilute acids.

If a metal reacts with a dilute acid, the products will be a salt and hydrogen. The general equation for this reaction is:

metal + **a**cid → **s**alt + **h**ydrogen

For example:

magnesium + sulfuric acid → magnesium sulfate + hydrogen

$$Mg + H_2SO_4 \rightarrow MgSO_4 + H_2$$

zinc + hydrochloric acid → zinc chloride + hydrogen

$$Zn + 2HCl \rightarrow ZnCl_2 + H_2$$

> **Reactivity series:** A list of elements (usually metals) arranged in order of chemical reactivity based on observations of their reactions. The most reactive element is at the top of the list.

> **Exam tip**
>
> This general equation can easily be remembered with the mnemonic **MASH**.

Neutralisation reaction: A reaction in which an acid is neutralised and a salt is produced.

Required practical

Investigate reactions between dilute hydrochloric and sulfuric acids and metals (e.g. magnesium, zinc and iron)

Method

1 Small quantities of dilute hydrochloric acid and sulfuric acid were added to separate test tubes.
2 One piece of magnesium was added to each test tube. Observations were recorded in a suitable table. If a gas was produced, it was tested using a burning splint.
3 Steps 1 to 2 were repeated with a different metal each time.

Results

Metal added	Observations with hydrochloric acid	Observations with sulfuric acid
Magnesium	Bubbles produced quickly. Test tube felt hot. Gas gave a squeaky pop sound when a burning splint was applied.	Bubbles produced extremely quickly. Test tube felt hot. Gas gave a squeaky pop sound when a burning splint was applied.
Zinc	Tiny bubbles were produced on the surface of the zinc quite quickly.	Tiny bubbles were produced on the surface of the zinc quickly.
Copper	No visible reaction.	No visible reaction.
Iron	A small number of tiny bubbles slowly formed on the surface of the iron.	A small number of tiny bubbles formed on the surface of the iron.

Based on these results, it was concluded that:
● The order of reactivity is magnesium > zinc > iron > copper.
● The gas produced each time was hydrogen, but it was only produced in sufficient quantity for a positive test result with the magnesium reaction.

Metals and water

REVISED

Iron and steel react slowly with water to produce rust (see page 64), but a small number of metals react with water more vigorously. In these reactions, hydrogen gas is produced as well as a solution of the metal hydroxide. The general equation for this reaction is:

metal + water → metal hydroxide + hydrogen

For example:

sodium + water → sodium hydroxide + hydrogen

$2Na + 2H_2O \rightarrow 2NaOH + H_2$

The higher up the metal is in the reactivity series, the more vigorous the reaction with water will be.

Metals and oxygen

Some metals are so reactive that they oxidise quickly in air. Examples include potassium, sodium and lithium. Other less reactive metals must be heated for oxidation to occur. These observations of reactivity in air allow us to put metals into a reactivity series. The general equation for a metal reacting with oxygen is:

metal + oxygen → metal oxide

For example:

copper + oxygen → copper oxide

$$2Cu + O_2 \rightarrow 2CuO$$

Revision activity

Make a deck of cards with the names of the metals within this section. Make another deck of cards with the names of acids on, as well as oxygen and water cards. Pick a card from each deck – one from the metal pile and one from the other pile. Will the two react? If so, what will the products be?

Exam tip

Remember that oxidation can be defined as when an atom loses electrons, or when a substance gains oxygen (see page 45). The opposite of oxidation is reduction. Reduction occurs when an atom gains electrons or a substance loses oxygen. Together, reduction and oxidation are called redox.

Metal displacement reactions

A more reactive element will displace (push out) a less reactive metal from a compound. This is called a **displacement reaction**. Observations of displacement reactions can be used to rank metals into a reactivity series.

Displacement reactions can occur between solids and also in aqueous solutions.

Displacement reactions between metals and metal oxides

Iron oxide powder can be reacted with aluminium powder in the thermite reaction. The resulting reaction is highly **exothermic** and produces molten iron, as well as a white smoke of aluminium oxide.

iron(III) oxide + aluminium → iron + aluminium oxide

$$Fe_2O_3 + 2Al \rightarrow 2Fe + Al_2O_3$$

In this example, the aluminium has displaced the iron from iron oxide, so aluminium must be more reactive than iron. The aluminium has been oxidised and the iron (in iron(III) oxide) has been reduced.

Displacement reaction: A reaction in which a more reactive element displaces (pushes out) a less reactive element from a compound and takes its place. The less reactive element is produced as one of the chemical products in the reaction.

Exothermic: A reaction that releases energy, usually as heat.

Displacement reactions between metals and aqueous solutions of metal salts

If a piece of zinc metal is placed into an aqueous solution of copper(II) sulfate, the zinc turns black and then red-brown. Meanwhile, the blue colour of the copper(II) sulfate solution will fade. This is because the zinc is more reactive than copper, so the zinc displaces the copper. This reaction can be seen below:

zinc + copper(II) sulfate → zinc sulfate + copper

$$Zn + CuSO_4 \rightarrow ZnSO_4 + Cu$$

Observations of displacement reactions in solution can be used to put metals into a reactivity series, as in the following example.

Exam practice answers and quick quizzes at **www.hoddereducation.co.uk/myrevisionnotesdownloads**

Aqueous solutions of metal salts were placed into different test tubes. Pieces of metal were then added to the test tubes. The student noted whether a displacement reaction occurred in each tube. The metals were not added to their own salt, e.g. copper was not added to copper nitrate.

	Copper metal	Silver metal	Magnesium metal	Zinc metal
Copper nitrate		No reaction	Copper is displaced	Copper is displaced
Silver nitrate	Silver is displaced		Silver is displaced	Silver is displaced
Magnesium nitrate	No reaction	No reaction		No reaction
Zinc nitrate	No reaction	No reaction	Zinc is displaced	
Number of reactions	1	0	3	2

- The number of reactions for each metal indicates how reactive it is. From these results we can conclude that the reactivity series for these four metals is:

magnesium > zinc > copper > silver

The reactivity series

REVISED

You need to know the order of reactivity of some common metals for the exam. Carbon and hydrogen have been included as well, even though they are non-metals. This is because carbon can be used to displace some metals from their oxide compounds (see pages 65–6), and because knowing where hydrogen is in the reactivity series tells you which metals will react with dilute acids.

potassium   most reactive
sodium
lithium
calcium
magnesium
aluminium
carbon
zinc
iron
hydrogen
copper
silver
gold   least reactive

Exam tip

It is a good idea to try to devise your own mnemonic for remembering this sequence. The sillier the better, because it makes it more likely it will stick in your head. For example: people say little children make all colourful zebras in Hawaii chew salty grass.

Now test yourself

TESTED

30 Write a word equation for the reaction of calcium with hydrochloric acid.
31 Write a balanced symbol equation for the reaction of zinc with sulfuric acid.
32 Write a balanced symbol equation for the reaction of magnesium with oxygen to form MgO.
33 Magnesium oxide powder is heated with tin powder. There is no reaction. What can be concluded about the reactivity of the two metals?

→

34 A student added small pieces of metals X, Y and Z to aqueous solutions of X chloride, Y chloride and Z chloride. The results are shown below. Deduce the order of reactivity.

	Metal X	Metal Y	Metal Z
X chloride solution		X displaced	X displaced
Y chloride solution	No reaction		No reaction
Z chloride solution	No reaction	Z displaced	

Answers on page 126

Rusting

Conditions needed for rusting

Rusting is a chemical reaction in which iron or steel is **oxidised** to form hydrated iron(III) oxide. Rusting is an example of a **corrosion** reaction. Many metals corrode, but only iron and steel rust. Remember that steel is an **alloy** of iron with carbon and sometimes other elements.

During rusting, the chemical equation is:

iron + oxygen → hydrated iron(III) oxide

For iron or steel to rust, the following conditions are required:
● Oxygen must be present.
● Water must be present.

Rusting occurs more quickly if dissolved ionic substances are present alongside the oxygen and water, for example acids or salts.

The conditions required for rusting to occur can be investigated using the method shown in Figure 2.9.

> **Typical mistake**
>
> Students often wrongly apply the term *rusting* to metals other than iron and steel.

> **Oxidation:** A process in which a substance gains oxygen (or loses electrons).
>
> **Corrosion:** A chemical reaction between a metal and substances in the environment (usually oxygen), which gradually destroys the metal.
>
> **Alloy:** A mixture of a metal element with another element (which is usually a metal, but can sometimes be carbon).

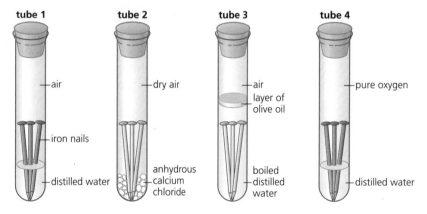

Figure 2.9 Investigating the conditions needed for rusting to occur

The nails in tubes 1 and 4 rust because oxygen and water are present. The nails in tube 2 do not rust because there is no water present (the calcium chloride absorbs any moisture in the air). The nails in tube 3 do not rust because there is no oxygen present (boiling the water removes any dissolved oxygen, and the oil and rubber stopper prevent any additional oxygen from the air dissolving into the water).

Preventing rusting

REVISED

Methods used to prevent rusting include barrier methods and chemical methods. Barrier methods work by preventing oxygen and water from reaching the surface of the iron or steel. Chemical methods use more reactive metals to prevent the iron or steel from oxidising.

Barrier methods

The surface of the iron or steel object can be painted, oiled, greased or coated in plastic or a less reactive metal in order to prevent oxygen and water from touching the iron or steel. Which method is used depends on the situation:

● The chain on a bicycle cannot be painted or coated in plastic because that would stop it from working correctly, so it is oiled or greased.
● The inside of a food can cannot be oiled or greased because that might make the food taste bad, so it is coated in plastic or a less reactive metal.
● The cheapest way to protect a fence or some railings for a long period of time is to paint them.

Chemical methods

A more reactive metal that is connected to an iron or steel object will become oxidised instead of the iron. While this oxidation is happening, it actually prevents the iron or steel from rusting. A block of zinc or magnesium can be bolted to the steel hull of a ship or the steel pillars of an oil rig to prevent the steel from rusting. This is called **sacrificial protection** because the more reactive metal will be lost through corrosion instead of the steel. However, it is easy to replace the blocks of the more reactive metal when they have corroded away. It would be very difficult to replace the steel hull or pillars.

Galvanising a steel object means coating it with a thin layer of zinc, which works as a barrier to oxygen and water. However, even if the zinc layer is scratched off and the surface of the steel is exposed, the remaining zinc will be oxidised instead of the steel because it is more reactive than iron. Therefore, the zinc will continue to prevent rusting through sacrificial protection.

Sacrificial protection: A method of preventing rusting which involves a more reactive metal being oxidised instead of the iron or steel.

Galvanising: Coating a steel object with a layer of zinc to prevent rusting.

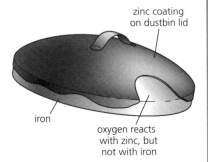

Figure 2.10 **Galvanising the steel dustbin lid with a layer of zinc prevents rusting**

Now test yourself

TESTED

35 State the conditions required for a steel object to rust.
36 Describe an experiment to investigate whether rusting is faster in sea water than fresh water.
37 Suggest a method of preventing the rusting of a steel shed.
38 Suggest a method of preventing the rusting of an iron pipe that runs along the bottom of a river.
39 Suggest a method of preventing the rusting of the moving parts of a farm machine.

Answers on page 126

Raw materials to metals

From rocks to metals

REVISED

Very unreactive metals, such as platinum, gold and silver, can be found in rocks as the metallic element itself. To obtain the metal, the rock is crushed and heated until the metal melts and can easily be separated.

Most metallic elements react with oxygen and can be found in rocks as compounds that contain the metal element bonded to oxygen (usually) or sometimes other elements such as sulfur and carbon. A rock that contains enough of a metal element or a metal compound to make it worth extracting the rock from the ground is called an **ore**.

Ore: A rock that contains enough of a metal or metal compound to make it economically worth extracting the rock from the ground.

Steps needed for extraction

The first step in extracting a metal is to mine or quarry the rock from the ground. The second step depends on the reactivity of the metal (see below).

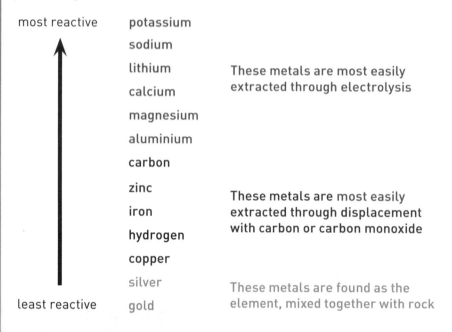

most reactive

potassium

sodium

lithium

calcium — These metals are most easily extracted through electrolysis

magnesium

aluminium

carbon

zinc

iron — These metals are most easily extracted through displacement with carbon or carbon monoxide

hydrogen

copper

silver

least reactive — gold — These metals are found as the element, mixed together with rock

Electrolysis could also be used to extract the metals that are below carbon in the reactivity series, but it is much more expensive than heating the metal oxide with carbon. Metals that are more reactive than carbon cannot be displaced by carbon, so electrolysis must be used.

After the metal has been obtained from its oxide, it usually needs to be purified. For example, copper is further purified using electrolysis, even though it is initially extracted using a displacement reaction.

Now test yourself

40 Name two elements that metals are often combined with when they are found in the Earth's crust.
41 State the meaning of the term 'ore'.
42 Explain why electrolysis is not used to extract iron from iron oxide.
43 Explain why calcium must be extracted by electrolysis and not by displacement using carbon.
44 Tin is more reactive than copper but less reactive than iron. Suggest how it could be extracted from tin oxide.

Answers on page 126

Extracting iron from iron ore – a case study

Overview of extraction

Iron is less reactive than carbon, so iron is extracted by heating iron oxide to a high temperature with carbon. The carbon acts as the reducing agent, and so reduces the iron oxide to iron. Limestone is also added during this process because it reacts with impurities in the iron ore, to remove them at the same time as the iron is produced.

The blast furnace

REVISED

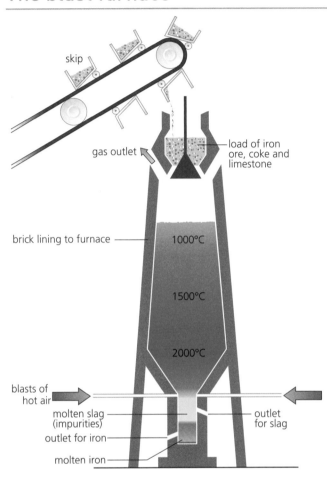

Figure 2.11 The blast furnace is used to extract iron from iron ore

The iron ore contains the compound iron(III) oxide, Fe_2O_3. This is added at the top of the blast furnace, together with limestone and coke (which is a cheap form of carbon). Hot air is blasted in near the bottom of the furnace. The carbon reacts with the oxygen in the air to form carbon monoxide. The carbon monoxide then reacts with the iron(III) oxide. The reaction is as follows:

iron(III) oxide + carbon monoxide → iron + carbon dioxide

$$Fe_2O_3 + 3CO \rightarrow 2Fe + 3CO_2$$

As the limestone reacts with impurities in the ore, it forms slag, which is a waste by-product from the process. The raw materials are added continually to the furnace, while the iron and slag are removed from the bottom of the furnace at intervals. The blast furnace runs 24 hours a day, 365 days a year because it is cheaper to keep it running at a very high temperature than it would be to allow it to cool and then heat it back up again.

Now test yourself

45 State the name of the reducing agent used in the blast furnace.
46 Why is limestone added to the blast furnace?
47 What is oxidised during the extraction of iron in the blast furnace?

Answers on page 126

TESTED

Extracting aluminium using electrolysis – a case study

Overview of extraction

Aluminium is more reactive than carbon, so it cannot be produced from its oxide ore through a displacement reaction with carbon as the reducing agent. Many years ago, aluminium used to be extracted from aluminium oxide in a displacement reaction, using sodium as the reducing agent. This was a very expensive and dangerous method and so for the last 100 years or so, aluminium has been extracted from aluminium oxide using electrolysis instead.

> **Exam tip**
>
> You do not need to be able to recall the details of these processes for your exam, but should be able to comment on a metal extraction process when you are given appropriate information in the exam.

Electrolysis of aluminium oxide

Aluminium oxide is purified from an ore called bauxite. In order for it to be electrolysed, aluminium oxide must either be melted or dissolved, so that the Al^{3+} and O^{2-} ions can move and are therefore able to conduct electricity. Unfortunately, aluminium oxide is insoluble in water, and it would require a temperature of $2072\,°C$ to melt it, which would be expensive and potentially dangerous.

However, scientists have discovered that aluminium oxide can be dissolved in another molten ionic compound called cryolite. Cryolite melts at approximately $1000\,°C$, which means that less energy is required to electrolyse a solution of aluminium oxide in molten cryolite than would be needed to electrolyse molten aluminium oxide at over $2000\,°C$.

Nonetheless, the process uses a lot of energy, both in heating and also in electrical energy to achieve electrolysis, so aluminium is relatively expensive to extract.

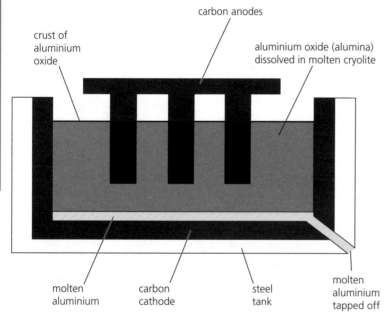

Figure 2.12 Electrolysis is used to extract aluminium from aluminium oxide

Now test yourself

48 Explain why carbon cannot be used to reduce aluminium in aluminium oxide.
49 Name the method used to extract aluminium from aluminium oxide.
50 Why is the aluminium oxide dissolved in cryolite rather than simply being melted?
51 Write a half equation to show an aluminium ion being reduced to form an aluminium atom.

Answers on page 126

TESTED

Using metals

The uses of a metal are explained by its properties

Most metals have similar **physical properties** to each other. For example, they have high melting points, they are malleable and they are good conductors of heat and electricity. However, there are still some large differences in these properties. For example, copper melts at $1085\,°C$, but tungsten melts at $3422\,°C$. The physical properties of a metal can usually be used to explain why it is chosen for a particular use.

> **Physical properties:**
> Descriptions of the ways that a substance characteristically behaves when not reacting with other chemicals. For example, its melting point, electrical conductivity, whether it is brittle, strong, hard, etc.

Aluminium

Aluminium has a low density compared to other metals and it resists corrosion because it forms a protective layer of aluminium oxide on its surface. This is why it is used for the body panels of luxury sports cars – it does not rust, and it allows them to accelerate faster by making the car lighter than it would be if another metal was used.

Aluminium is also used for:
- drinks cans, because it does not react with the acids in the drink
- high voltage power lines, because it is less dense than copper. Even though it is not such a good conductor, the pylons do not need to be as strong (and therefore as expensive) as they would need to be to hold very heavy copper cables.

Copper

Copper is an excellent conductor of electricity, which explains why it is used in domestic electrical wiring and in electronic devices. It is also used in plumbing because it does not react very quickly with water.

Iron and steel

Iron is rarely used as a pure metal because its properties are significantly improved if it is alloyed to make steel (steels are harder and stronger than pure iron). There are many types of steel, depending on the other elements that are mixed with the iron.

Table 2.8 A comparison of three types of steel

Type of steel	Contains	Properties	Uses
Low-carbon steel (mild steel)	< 0.3% carbon	Malleable	Car body panels
High-carbon steel	1–2% carbon	Very strong	Springs, cables
Stainless steel	> 10% chromium	Resists corrosion	Cutlery, medical instruments, tools, sinks

> **Exam tip**
>
> You should be familiar with the properties and uses of low-carbon steel, high-carbon steel and stainless steel because you could be asked about this in the exam.

> **Revision activity**
>
> Copy out Table 2.8 but leave the final three columns blank (apart from the headers). Try filling it in correctly and then check to see how well you did.

52 Suggest why stainless steel's physical properties make it suitable for use as cutlery.
53 Why are saucepans often made from aluminium, and rarely made from copper?
54 What is meant by the term 'alloy'?
55 Why are domestic wires made from copper but high voltage power lines made from aluminium?

Answers on page 126

Acids and their properties

The pH scale

REVISED

The pH scale is used to describe how **acidic** or **alkaline** a solution is. The most common acids and alkalis lie between pH 0 and pH 14. pH values below 7 are acidic; pH 7 is neutral; pH values above 7 are alkaline.

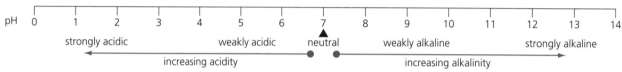

Figure 2.13 The pH scale is used to describe whether a solution is acidic, neutral or alkaline

Acid: A substance that reacts by giving away a hydrogen ion (H^+) in a reaction. Acids produce salts when they react, and their solutions have a pH less than 7.

Alkali: A base that dissolves in water. An alkaline solution has a pH greater than 7.

Testing for acids

REVISED

Acids can be identified using an **indicator**, which will show a different colour depending on the pH of a solution.

Indicator: A substance that changes colour depending on how acidic or alkaline a solution is.

Litmus is an indicator which shows whether a solution is acidic or alkaline. Universal indicator is useful because it has a large number of colours which can provide an approximate pH value when compared with a reference colour scale.

Litmus changes colour at pH 7. Other indicators change colour at different pH values. Figure 2.15 shows the colours of methyl orange and phenolphthalein indicators at different pH values.

Exam tip

The indicators you need to know about are litmus, phenolphthalein, methyl orange and universal indicator. In preparation for the exam you should learn what colour they are in solutions of different pHs.

Figure 2.14 The colours shown by litmus (top) and universal indicator (bottom) at different pH values

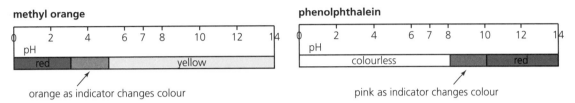

orange as indicator changes colour

pink as indicator changes colour

Figure 2.15 The colours shown by methyl orange (left) and phenolphthalein (right) at different pH values

Reactions of acids

REVISED ☐

There are several different ways to neutralise an acid, but all of them produce a salt as one of the products.

Acid plus metal

Dilute solutions of acids will react with any metal that is above hydrogen in the reactivity series. The products of these reactions are a metal **salt** and hydrogen gas. The test for hydrogen is described on page 80.

The general equation for this reaction is:

metal + acid → salt + hydrogen

For example:

calcium + hydrochloric acid → calcium chloride + hydrogen

$$Ca \quad + \quad 2HCl \quad \rightarrow \quad CaCl_2 \quad + \quad H_2$$

Acid plus metal oxide

Acids react with metal oxides (which are often called **bases**) to produce a salt plus water. The general equation for this reaction is:

metal oxide + acid → salt + water

For example:

magnesium oxide + nitric acid → magnesium nitrate + water

$$MgO \quad + \quad 2HNO_3 \rightarrow \quad Mg(NO_3)_2 \quad + H_2O$$

Acid plus metal hydroxide

Metal hydroxides are called alkalis if they are soluble (for example, all group 1 hydroxides, like sodium hydroxide) or bases if they are insoluble (for example, zinc hydroxide). Acids react with alkalis and insoluble metal hydroxides to produce a salt plus water. The general equation for this reaction is:

metal hydroxide + acid → salt + water

Because soluble hydroxide compounds are alkalis, this is often written as:

alkali + acid → salt + water

For example:

potassium hydroxide + sulfuric acid → potassium sulfate + water

$$2KOH \quad + \quad H_2SO_4 \rightarrow \quad K_2SO_4 \quad + 2H_2O$$

Salt: A compound that is produced when an acid is neutralised, which contains the negative ion from the acid.

Exam tip

This general equation can easily be remembered with the mnemonic **MASH**.

Exam tip

Remember that the acid used determines the type of salt made during a neutralisation reaction. Hydrochloric acid always produces chloride salts, which contain the chloride ion, Cl^-. Nitric acid always produces nitrate salts, which contain the nitrate ion, NO_3^-.

Base: A substance that reacts with an acid, accepting a hydrogen ion (H^+) in a reaction.

Acid plus metal carbonate

Acids react with metal carbonates to produce a salt, water and carbon dioxide gas. The general equation for this reaction is:

metal carbonate + acid → salt + water + carbon dioxide

For example:

calcium carbonate + hydrochloric acid → calcium chloride + water + carbon dioxide

$$CaCO_3 \quad + \quad 2HCl \quad → \quad CaCl_2 \quad + H_2O + \quad CO_2$$

copper carbonate + nitric acid → copper nitrate + water + carbon dioxide

$$CuCO_3 \quad + \quad 2HNO_3 → \quad Cu(NO_3)_2 \quad + H_2O + \quad CO_2$$

Ions in acids and alkalis

Aqueous solutions of acids contain hydrogen ions, H^+. When acidic solutions react, they give away these hydrogen ions. Aqueous solutions of alkalis contain hydroxide ions, OH^-. This means that an acid–alkali neutralisation reaction can be summarised by the following ionic equation:

$$H^+ + OH^- → H_2O$$

The other ions present during a neutralisation reaction are what make up the salt. For example, if the acid was hydrochloric acid, the negative ion would be chloride, Cl^-. If the alkali was sodium hydroxide, the positive ion would be sodium, Na^+. As these two ions are present in the solution, when the water is evaporated they form crystals of sodium chloride, $NaCl$.

Now test yourself

TESTED

56 State the colour of phenolphthalein at the following pHs: 3, 6, 7, 9, 12.
57 Write a word equation for the reaction of magnesium and sulfuric acid.
58 Write a word equation for the reaction of zinc oxide with hydrochloric acid.
59 Write a symbol equation for the reaction of lithium hydroxide with nitric acid.

Answers on page 127

Revision activity

Ask someone to test your ability to either remember or work out the products for any of the equations in this section, including the general equations, word equations and symbol equations.

Acid–alkali titrations

What is a titration?

REVISED

A titration is an experiment which allows you to measure the exact volumes of an acid and an alkali that are needed to produce a neutral solution. If you know the exact concentration of one of the solutions, you can work out the concentration of the other solution.

Required practical

Carrying out an acid–alkali titration

Aim

To determine the volume of hydrochloric acid (of an unknown concentration) that is required to neutralise 25.0 cm³ of sodium hydroxide which has a concentration of 0.1 mol/dm³.

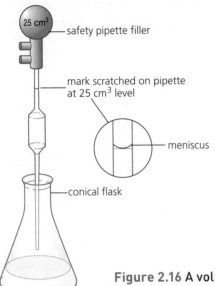

Volumetric pipette: A piece of apparatus that is used to suck up a specific volume of a liquid in the first stage of a titration.

Pipette filler: A piece of apparatus that fits on the top of a volumetric pipette to provide suction in a safe way.

Burette: A piece of apparatus used in a titration that releases a liquid into the conical flask via a tap at the bottom. A burette will accurately measure the volume of liquid that has been added.

Figure 2.16 A volumetric pipette is used to add exactly 25.0 cm³ of sodium hydroxide into a conical flask

Method

1. A **volumetric pipette** was fitted with a safety **pipette filler** and used to measure exactly 25.0 cm³ of the alkali into a conical flask, which was placed on a white tile.
2. Two drops of a suitable indicator, in this case phenolphthalein, were added to the conical flask. Phenolphthalein will turn pink or red in the alkali.
3. A **burette** was filled with the acid so that the liquid's **meniscus** was at the zero mark.
4. Acid was released slowly from the burette, a little at a time, while the conical flask was swirled. When the indicator started to fade, the acid was added even more slowly – a drop at a time.
5. When the indicator first turned permanently colourless, the titration was considered over. The volume of the acid used to cause this loss of colour was recorded.
6. The process was repeated until two of the results were **concordant**.

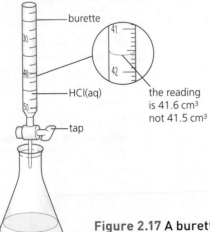

Meniscus: The surface tension (or 'skin') of a liquid. At the edges of a container, such as a beaker, pipette or burette, the meniscus appears to climb up the glass. Measurements are always taken from the lowest (middle) point on the meniscus.

Concordant: Results which are very close together, e.g. within 0.2 cm³. In some situations, concordant results are required to be within 0.1 cm³.

Figure 2.17 A burette is used to add the acid into the conical flask during a titration

Analysis

- The first titration result was counted as a rough value, so it was ignored (which is usual practice).
- The mean volume of the acid was calculated from the two results which were concordant.

Now test yourself

60 Name the piece of apparatus used to measure 25.0 cm³ of alkali during a titration.
61 Name the piece of apparatus used to suck the liquid into a pipette during a titration.
62 Why is a white tile placed under the conical flask in a titration?
63 When do you stop adding the chemical from the burette in an acid–alkali titration?

Answers on page 127

Typical mistake

Do not confuse a volumetric pipette with a dropper pipette. A volumetric pipette is used to accurately measure a specific volume of a liquid. A dropper pipette is used to transfer small volumes (i.e. drops) of one liquid to another container.

Bases and alkalis

Proton transfer in neutralisation reactions

REVISED

Bases are substances which neutralise **acids**. Because acids are proton (H⁺) donors, a base is a chemical which accepts a proton in a reaction. Bases are therefore proton acceptors.

If a base dissolves in water, it is an **alkali**. Alkaline solutions contain hydroxide ions, OH⁻. These hydroxide ions react with the hydrogen ions (protons) in an acidic solution, so an acid–alkali neutralisation reaction can be summarised with the following ionic equation:

$$H^+ + OH^- \rightarrow H_2O$$

Base: A substance that reacts with an acid, accepting a hydrogen ion (H⁺) in a reaction.

Acid: A substance that reacts by giving away a hydrogen ion (H⁺) in a reaction. Acids produce salts when they react, and their solutions have a pH less than 7.

Alkali: A base that dissolves in water. An alkaline solution has a pH greater than 7.

Ammonia acts as a base

REVISED

Many bases are metal compounds, such as metal oxides, hydroxides and carbonates. Ammonia (NH₃) is a compound made from two non-metals, and it is also an important base. Ammonia has a simple molecular covalent structure (see page 36) so it has a very low boiling point. It is also a very soluble gas, and when it dissolves in water it produces a solution of ammonium hydroxide, NH₄OH. This solution contains hydroxide ions and has a pH which is greater than 7.

The following equations show how ammonia and ammonia solution (NH₄OH) both act as a base:

ammonia + hydrochloric acid → ammonium chloride

$$NH_3 + HCl \rightarrow NH_4Cl$$

ammonium hydroxide + hydrochloric acid → ammonium chloride + water

$$NH_4OH + HCl \rightarrow NH_4Cl + H_2O$$

ammonium hydroxide + sulfuric acid → ammonium sulfate + water

$$2NH_4OH + H_2SO_4 \rightarrow (NH_4)_2SO_4 + 2H_2O$$

Now test yourself

64 Which ion is present in all acidic solutions?
65 What is the difference between a base and an alkali?
66 Which ion is present in all alkaline solutions?
67 What is the definition of an acid?

Answers on page 127

Salts

What is a salt?

A salt is an ionic compound which is produced when an acid is neutralised by a base. The positive ion present in a salt comes from the base, and the negative ion comes from the acid. Because salts are ionic compounds, they have predictable properties (see also page 39). These properties include:

- high melting points
- not being able to conduct electricity when solid
- being able to conduct electricity when molten or dissolved
- solubility in water (for most salts, although some are not – see below).

To make a specific salt, you can usually select an acid and a base that contain the desired ions. For example, to make copper nitrate, you need nitric acid to provide the nitrate ion. The copper ion could come from copper oxide. The copper ion could not come from copper metal as it is not reactive enough to react with dilute acids.

Solubility rules for salts and other ionic compounds

You need to be able to recall and apply solubility rules to work out if a given ionic compound is soluble or insoluble in water. In the rules that follow, green text indicates what is soluble and red text indicates what is insoluble.

- All common **sodium**, **potassium** and **ammonium** compounds are soluble.
- All **nitrates** are soluble.
- All common **chlorides** are soluble, except for **silver chloride** and **lead(II) chloride**.
- Most **sulfates** are soluble, except for **barium sulfate**, **calcium sulfate** and **lead(II) sulfate**.
- Most **carbonates** are insoluble, except for **sodium**, **potassium** and **ammonium**.
- Most **hydroxides** are insoluble, except for **sodium**, **potassium** and **ammonium**. You also need to know that calcium hydroxide is slightly soluble.

Example

Deduce whether lead nitrate is soluble or insoluble.
- All nitrates are soluble, and this includes lead nitrate.

Revision activity

Place the names of some common positive ions (metals plus ammonium) onto small sticky notes. Place the names of some common negative ions (e.g. carbonate, chloride, nitrate, sulfate, hydroxide) onto another set of sticky notes. Take one positive and one negative ion to make different compounds, e.g. copper carbonate. Using your knowledge of the solubility rules, deduce whether the compound you have created is soluble or not. Now check your deduction against the rules.

Preparing an insoluble salt

An insoluble ionic compound can be made using a **precipitation reaction**. Two soluble compounds are chosen as the reactants, with the desired positive ion present in one solution, and the desired negative ion present in the other solution. The desired compound forms as a solid and settles out.

> **Precipitation reaction:** A reaction that takes place when two solutions are mixed and one of the products created is an insoluble solid. The solid product forms as a precipitate and falls out of the solution.

Exam tip

When asked to suggest two soluble compounds to mix to produce a precipitate of an insoluble salt, choose a nitrate compound as the source of the metal ion because all nitrates are soluble. Since almost all sodium salts are soluble, choose a sodium compound as the source of the negative ion.

Required practical

Prepare a sample of pure, dry lead(II) sulfate

Method

1 $20\,cm^3$ of $0.1\,mol/dm^3$ lead(II) nitrate solution was added to $20\,cm^3$ of $0.1\,mol/dm^3$ sodium sulfate solution.
2 Solid lead(II) sulfate precipitated out, leaving a solution of sodium sulfate as the by-product.
3 The mixture was filtered. The lead(II) sulfate residue was washed with distilled water to remove any soluble ions.
4 The residue was dried either by pressing between paper towels, or being placed in a warm place such as a drying oven.

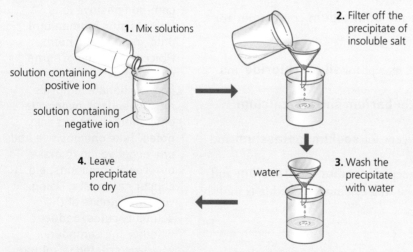

1. Mix solutions

solution containing positive ion

solution containing negative ion

2. Filter off the precipitate of insoluble salt

4. Leave precipitate to dry

water

3. Wash the precipitate with water

Figure 2.18 Preparing a sample of an insoluble salt using a precipitation reaction

Now test yourself

68 Work out whether potassium hydroxide is soluble or insoluble.
69 Work out whether calcium carbonate is soluble or insoluble.
70 Work out whether cobalt chloride is soluble or insoluble.
71 Describe how you would prepare a pure, dry sample of the insoluble compound cobalt carbonate. Name the reactants you would use.

Answers on page 127

Preparing soluble salts

Using an insoluble base

A soluble salt can sometimes be prepared using an insoluble base. This is usually more straightforward than using a soluble base, because the base can be added until it is in **excess**, then the excess base can be filtered out and disposed of. Insoluble bases include metals, most metal oxides and most metal carbonates.

The desired negative ion is obtained from the appropriate acid, and the positive ion is obtained from the insoluble base. For example, to make copper(II) sulfate, the required acid is sulfuric acid and the copper ion could come from copper oxide or copper carbonate (both of which are insoluble).

> **Excess:** A reactant is in excess if there is some left unreacted at the end of the reaction. The other reactant is described as the **limiting** reactant.

Required practical

Prepare a sample of pure, dry hydrated copper(II) sulfate crystals starting from copper(II) oxide

Method

1 20 cm^3 of dilute sulfuric acid was warmed gently in a beaker to approximately 50 °C.
2 One spatula of copper oxide was added and the mixture was stirred with a glass rod until the copper oxide had completely reacted and no powder was left in the beaker.
3 Step 2 was repeated until some copper oxide remained in the beaker.
4 The mixture was filtered and the residue was discarded.
5 The filtrate was poured into an evaporating basin and heated until approximately half the solution had evaporated.
6 The solution was left to cool and crystallise.
7 The crystals were removed using tweezers and dried between paper towels (alternatively they could have been left in a warm place).

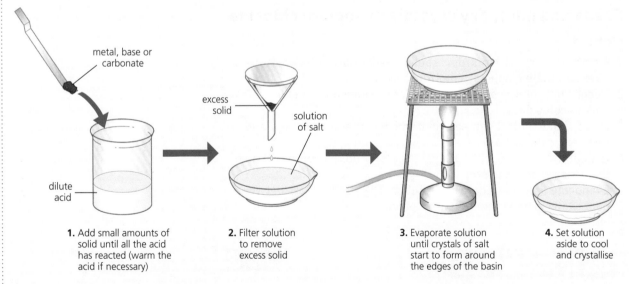

metal, base or carbonate

excess solid

solution of salt

dilute acid

1. Add small amounts of solid until all the acid has reacted (warm the acid if necessary)

2. Filter solution to remove excess solid

3. Evaporate solution until crystals of salt start to form around the edges of the basin

4. Set solution aside to cool and crystallise

Figure 2.19 Preparing a soluble salt using an insoluble base

Using a soluble base

If the chosen acid and base are both soluble, it is difficult to tell when the acid has been used up and the solution has been neutralised. One method involves taking samples of the reaction mixture and spotting them onto a suitable indicator paper. The pH of the mixture as shown by the indicator paper will tell you whether you need to add more of the acid or alkali. When the reaction mixture is neutral, you can evaporate half the water and allow what is left to crystallise, as described in Figure 2.20.

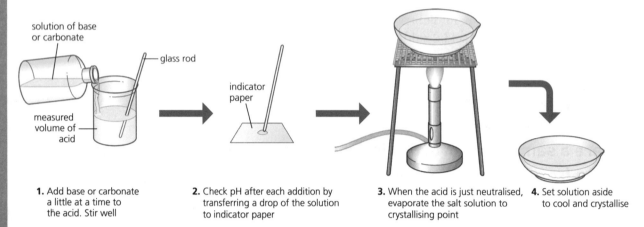

1. Add base or carbonate a little at a time to the acid. Stir well

2. Check pH after each addition by transferring a drop of the solution to indicator paper

3. When the acid is just neutralised, evaporate the salt solution to crystallising point

4. Set solution aside to cool and crystallise

Figure 2.20 **Preparing a soluble salt using a soluble base**

An alternative method using two titrations is outlined below. The first titration allows you to work out the volumes needed for a neutral solution using an indicator. The second titration does not use an indicator because you know the required volumes from the first titration.

Required practical

Preparing pure, dry crystals of sodium chloride
Method

1 Titrate $25.0\,cm^3$ of $1\,mol/dm^3$ sodium hydroxide with two drops of phenolphthalein against $1\,mol/dm^3$ hydrochloric acid.
2 Record the volume of acid needed for the indicator to change from pink to colourless.
3 Discard the neutral salt solution, which contains indicator, and rinse the conical flask.
4 Repeat the titration without the indicator. Add the volume of acid that was recorded in step 2.
5 Evaporate and crystallise the solution. Dry the crystals between paper towels and then in a warm place.

Now test yourself

72 Describe how to prepare a pure, dry sample of calcium chloride, starting from the insoluble base calcium carbonate.
73 Describe how to prepare a pure, dry sample of potassium nitrate.

Answers on page 127

Tests for ions, gases and water

Flame tests

REVISED

Flame tests can be used to identify the presence of certain metal **cations** in a solid or solution.

Cation: A positively charged ion.

1 A nichrome wire loop was dipped into concentrated hydrochloric acid, then heated in a blue Bunsen flame until it gave no further colour to the flame.
2 The wire loop was then dipped into the solid ionic compound or the solution to be tested, then held in the blue Bunsen flame.
3 The colour of the flame was then recorded.

Table 2.9 The flame colours caused by certain metal ions

Ion present	Flame colour
Li^+	Red
Na^+	Yellow
K^+	Lilac
Ca^{2+}	Orange-red
Cu^{2+}	Blue-green

Testing for positive ions using sodium hydroxide

REVISED

Sodium hydroxide can be used to test for both ammonium ions and several different transition metal ions in aqueous solutions.

Ammonium cations (NH_4^+) in a solution

1 Sodium hydroxide solution was added to a test tube containing the solution to be tested.
2 The mixture was warmed and some damp red litmus paper was held near the top of the test tube.
3 If ammonia gas is produced, the damp red litmus paper will turn blue.

Transition metal cations in a solution

1 A few drops of sodium hydroxide were added to a test tube containing the solution to be tested.
2 The colour of the precipitate formed was recorded.

Table 2.10 The precipitate colours caused by certain metal ions with sodium hydroxide

Ion present	Precipitate colour	Ionic equation
Cu^{2+}	Blue	$Cu^{2+} + 2OH^- \rightarrow Cu(OH)_2$
Fe^{2+}	Green	$Fe^{2+} + 2OH^- \rightarrow Fe(OH)_2$
Fe^{3+}	Brown	$Fe^{3+} + 3OH^- \rightarrow Fe(OH)_3$

Testing for negative ions in a solution

REVISED

Several **anions** can be identified in aqueous solutions using precipitation reactions, including carbonate ions, sulfate ions and halide ions.

Anion: A negatively charged ion.

Carbonate ions, CO_3^{2-}

1 A small volume of dilute hydrochloric acid was added to a test tube containing the solution to be tested.
2 A gas was produced, and this gas was bubbled through limewater.
3 The limewater turned cloudy, which was a positive result for carbon dioxide, and hence the original solution contained carbonate ions.

Sulfate ions, $SO_4{}^{2-}$

1 A squirt of dilute hydrochloric acid was added to a test tube containing the solution to be tested.
2 A few drops of barium chloride solution were added to the test tube.
3 A positive result can be seen as a white precipitate of barium sulfate:
   $Ba^{2+}(aq) + SO_4{}^{2-}(aq) \rightarrow BaSO_4(s)$.

Halide ions

Halide ions are the simple negative ions formed when the halogen atoms in group 7 have gained an electron, e.g. chloride (Cl^-), bromide (Br^-), and iodide (I^-).

1 A squirt of dilute nitric acid was added to a test tube containing the solution to be tested.
2 A few drops of silver nitrate solution was added to the test tube.

Table 2.11 The precipitate colours caused by halide ions with silver nitrate

Ion present	Precipitate colour	Ionic equation
Cl^-	White	$Ag^+ + Cl^- \rightarrow AgCl$
Br^-	Cream	$Ag^+ + Br^- \rightarrow AgBr$
I^-	Yellow	$Ag^+ + I^- \rightarrow AgI$

> **Revision activity**
>
> Make separate flash cards for each ion, method and positive result in this section. Shuffle them and then see if you can match them up correctly.

> **Exam tip**
>
> It can be helpful to remember the colours of the silver halide precipitates using: 'milk – cream – butter'.

Testing for gases

REVISED

Simple laboratory tests can be used to identify different gases.

Hydrogen

1 The gas was collected in a test tube.
2 A burning splint was brought near to the top of the test tube.
3 A positive result for hydrogen is a small explosion with a squeaky pop sound.

Oxygen

1 The gas was collected in a test tube.
2 A glowing splint was inserted into the test tube.
3 A positive result for oxygen is if the splint relights.

Carbon dioxide

1 The gas was bubbled through limewater, or shaken with limewater in a stoppered test tube.
2 A positive result for carbon dioxide is if the limewater turns cloudy.

Ammonia

1 The gas was collected in a test tube.
2 A piece of damp red litmus paper, or damp universal indicator paper, was held near to the top of the test tube.
3 A positive result for ammonia is if the indicator paper turns blue or purple.

Chlorine

1 The gas was collected in a test tube.
2 A piece of damp litmus paper, or damp universal indicator paper, was held near to the top of the test tube.
3 A positive result for chlorine is if the indicator paper is bleached white.

Water can be detected using anhydrous copper(II) sulfate because it turns the sulfate from white to blue.

To test to see if a sample of water is pure, you can measure its boiling point as pure water boils at exactly 100 °C. After the water has completely evaporated, there should be no solid residue left behind.

Now test yourself

74 A white compound was dissolved in distilled water. A flame test gave a lilac flame. When a separate sample was tested using dilute hydrochloric acid and then barium chloride, a white precipitate was produced. Identify the ions present and use this to state the name of the compound.

75 A mystery solution was tested with sodium hydroxide solution and a blue precipitate was produced. A separate sample was tested with nitric acid and then silver nitrate, which produced a cream precipitate. Identify the ions present and use this to state the name of the compound.

76 Describe how to test a gas to determine whether it is hydrogen or oxygen.

77 Describe a single test to determine whether a sample of a gas is chlorine or ammonia, and state the positive result for each gas.

Answers on page 127

Summary

- The alkali metals are in group 1. They react quickly with water to produce an alkaline solution and hydrogen. They get more reactive further down the group.
- The halogens are in group 7. They get less reactive further down the group. Their melting and boiling points increase further down the group.
- Oxidation is defined as when a substance gains oxygen or loses electrons in a reaction. Reduction is defined as losing oxygen or gaining electrons. Reduction and oxidation occur simultaneously and this is called redox.
- The air is 78% nitrogen and 21% oxygen. The remaining 1% is mostly argon with a small amount of carbon dioxide.
- Metals can be placed into a reactivity series by observing their reactions with oxygen, acids and metal salt solutions (displacement reactions).
- Rusting is the oxidation of iron or steel, and it requires oxygen and water.
- Metals can be extracted from their oxide ores by displacement using carbon if they are below carbon in the reactivity series.
- Metals that are more reactive than carbon must be extracted using electrolysis.
- How we use metals depends on their properties.

- Acids donate H^+ ions when they react and are called proton donors. They have a pH which is below 7.
- Bases accept H^+ ions when they neutralise acids and are called proton acceptors. If a base dissolves in water, it forms an alkaline solution which contains OH^- ions and has a pH greater than 7.
- Acid + metal → salt + hydrogen
- Acid + metal oxide or hydroxide → salt + water
- Acid + metal carbonate → salt + water + carbon dioxide
- A titration can be used to measure the volume of an acid that is required to neutralise a specific volume of an alkali.
- Solubility rules can be applied to make predictions about whether a solid ionic compound will dissolve in water.
- Soluble salts can be prepared in solution and then crystallised.
- Insoluble salts can be prepared using precipitation reactions.
- Positive ions in a solution can be detected using flame tests or by adding NaOH.
- Negative ions can be detected by precipitation reactions.
- Gases can be identified using specific tests.
- Pure water turns anhydrous copper(II) sulfate from white to blue, and has a boiling point of exactly 100 °C.

Exam practice

1 The elements in group 1 are called the alkali metals. A teacher demonstrated the reaction of sodium with water by placing a very small piece of sodium into a large glass container that had water in the bottom.

   a Describe three observations that the students in the class would be able to make. [3]

   b Write a word equation for the reaction of sodium with water. Identify the product which makes the solution alkaline. [3]

   c Explain why the reaction of lithium with water is similar to the reaction observed during the demonstration. [1]

   d Explain why the reaction between water and lithium is less vigorous than the reaction between water and sodium. [2]

   e Solutions made during these reactions contain the positive ion of the alkali metal which was used, for example Li^+, Na^+ or K^+. Describe how you could do a simple test to distinguish between separate solutions that contained one of these three ions. You should include the positive result for each ion in your answer. [6]

2 This question is about the halogens chlorine, bromine and iodine.

   a State the group number of the halogens. [1]

   b Copy and complete the table to show the appearance and state of chlorine, bromine and iodine at room temperature and pressure. [6]

Halogen	State at room temperature and pressure	Colour
Chlorine		
Bromine		
Iodine		

   c Chlorine will react violently with sodium to produce sodium chloride, NaCl. Write a balanced symbol equation for this reaction. [2]

   d The trend in reactivity of the halogens can be demonstrated using displacement reactions. In these reactions, a more reactive halogen in an aqueous solution will displace a less reactive halogen from a solution of its potassium salt.

      Copy and complete the table using ticks and crosses to show which reactions result in a reaction and which do not. [4]

	Chlorine solution	Bromine solution	Iodine solution
Potassium chloride solution	Reaction not investigated	✗	
Potassium bromide solution		Reaction not investigated	
Potassium iodide solution	✓		Reaction not investigated

   e Fluorine is the most reactive halogen element. If it is mixed with a solution of sodium bromide, a displacement reaction occurs. Write a word equation for the reaction of fluorine with sodium bromide. [1]

   f Explain why fluorine is more reactive than bromine. You should use ideas about electrons in your answer. [3]

   g During an experiment similar to the one described in part d, a student mixed up unlabelled solutions of sodium chloride, sodium bromide and sodium iodide. Describe a test that the student could do to correctly identify which solution was which. You should include the expected observations for each solution in your answer. [5]

3 The atmosphere is a mixture of a number of gases, some of which are elements and some of which are compounds.

a Copy and complete the table to identify the gases present in dry air, and their percentage proportions in the atmosphere. [4]

Gas	Percentage of dry air
	78
Oxygen	
	0.9
	0.04

b Describe an experiment using a metal that would allow you to determine the approximate percentage by volume of oxygen in air. [6]

c Carbon dioxide is a greenhouse gas. Explain what is meant by the term 'greenhouse gas'. [2]

d Carbon dioxide can be produced when copper(II) carbonate ($CuCO_3$) is heated strongly, causing it to break down and produce copper(II) oxide (CuO).

  i State the name given to this type of reaction. [1]

  ii Write a balanced symbol equation for this reaction. [2]

  iii Describe a test to identify whether a sample of a gas is carbon dioxide. Include the positive result in your answer. [2]

4 Observing how metals and some specific non-metals react with other substances allows them to be placed into a reactivity series, which lists the elements in order of reactivity, with the most reactive at the top. Part of a reactivity series is shown below:

magnesium
carbon
zinc
iron
lead
copper

a A student investigated the displacement reactions of four metals from the list given above. She added metals W, X, Y and Z to the nitrate salts of all of the metal elements in the list. The table shows the results recorded by the student.

	Metal W	Metal X	Metal Y	Metal Z
Magnesium nitrate solution	No visible reaction	No visible reaction	No visible reaction	No visible reaction
Zinc nitrate solution	No visible reaction	The surface of the metal turned black	No visible reaction	No visible reaction
Iron nitrate solution	No visible reaction	The surface of the metal turned black	The surface of the metal turned black	No visible reaction
Lead nitrate solution	No visible reaction	The surface of the metal turned grey	The surface of the metal turned grey	The surface of the metal turned grey
Copper nitrate solution	No visible reaction	The surface of the metal turned brown	The surface of the metal turned brown	The surface of the metal turned brown

  i Place the metals W, X, Y and Z in order of reactivity, with the most reactive at the top of the list. [1]

  ii Identify which metal was copper. Explain your answer. [2]

  iii Identify which metal was magnesium. Explain your answer. [2]

  iv Identify which metal from the reactivity series given above was not used in the experiment. Explain your answer. [2]

b The method of extracting a metal from its oxide ore depends on the reactivity of the metal. Choose a **metal** from the reactivity series provided that:

  i must be extracted using electrolysis [1]
    magnesium   **zinc**   **iron**   **copper**

  ii **cannot** be extracted by heating its oxide ore with carbon. [1]
    lead   **zinc**   **iron**   **magnesium**

Answers and quick quizzes online

ONLINE

Energy changes and enthalpy changes

Exothermic and endothermic reactions

Most chemical reactions release energy into the surroundings. These reactions are called **exothermic** reactions, and the temperature of the surroundings (or the solution that the reaction occurs in) increases. These reactions will feel hot to the touch. Examples of exothermic reactions include combustion, respiration and the reactions of alkali metals with water.

A smaller number of reactions absorb energy from the surroundings. These are called **endothermic** reactions, and the temperature of the surroundings in these examples decreases. Examples of endothermic reactions include photosynthesis and dissolving ammonium nitrate in water.

> **Exothermic:** A process which releases energy (usually as heat) from the chemicals into the surroundings.
>
> **Endothermic:** A process which absorbs energy (usually as heat) from the surroundings into the chemicals.

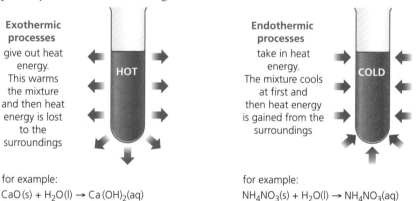

Exothermic processes give out heat energy. This warms the mixture and then heat energy is lost to the surroundings

HOT

Endothermic processes take in heat energy. The mixture cools at first and then heat energy is gained from the surroundings

COLD

for example:
$CaO(s) + H_2O(l) \rightarrow Ca(OH)_2(aq)$

for example:
$NH_4NO_3(s) + H_2O(l) \rightarrow NH_4NO_3(aq)$

Figure 3.1 Exothermic and endothermic reactions

> **Exam tip**
>
> It can help to remember that in an **exo**thermic reaction, heat energy **exits** the chemicals into the surroundings. In an **endo**thermic reaction, heat energy goes **into** the chemicals.

Enthalpy changes and ΔH

The amount of heat energy transferred in a chemical reaction is called the **enthalpy change**. Enthalpy change is shown by the symbol ΔH and it uses the units of **kilojoules per mole**, or kJ/mol. Because we are interested in the energy that has been lost or gained by the **chemicals** (as opposed to what happens to the surroundings), when the chemicals **lose** energy in an exothermic reaction, we say that the enthalpy change is **negative**. An endothermic reaction has a **positive** value for ΔH.

> **Enthalpy change, ΔH:** The heat energy change per mole of the **limiting reactant**. If the enthalpy change is negative, the reaction is exothermic.
>
> **Limiting reactant:** The reactant that runs out in a chemical reaction and therefore determines (or limits) when the reaction stops.

Energy level diagrams

Energy level diagrams can be used to show exothermic and endothermic reactions. If the products created have less stored chemical energy than the reactants, the reaction is exothermic, because energy must have been released.

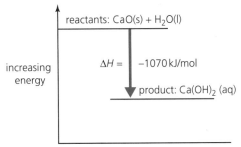

Figure 3.2 Energy level diagram for an exothermic reaction. Note that ΔH is negative

If the products have more stored chemical energy than the reactants, the reaction is endothermic, because energy has been absorbed.

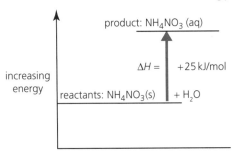

Figure 3.3 Energy level diagram for an endothermic reaction. Note that ΔH is positive

> **Revision activity**
>
> Copy out Figures 3.2 and 3.3 with none of the words. Revisit the blank diagrams later to see if you can remember all the words/labels and then check how you did.

Now test yourself

1 Single-use ice packs can be used for treating sports injuries. Suggest whether an exothermic or endothermic reaction would be needed in a single-use ice pack.
2 During a reaction between magnesium and sulfuric acid, the temperature in the boiling tube increased. Was this reaction exothermic or endothermic?
3 The combustion of methane is an exothermic reaction. Is the value of ΔH positive or negative?

Answers on page 127

Measuring enthalpy changes

Required practical

Investigate temperature changes accompanying combustion reactions

Figure 3.4 shows the apparatus used to measure the enthalpy of combustion of a liquid fuel.

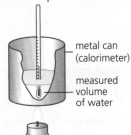

metal can (calorimeter)

measured volume of water

liquid burner — ethanol

Figure 3.4 The apparatus used to measure the enthalpy of combustion of a liquid fuel →

Method

1 100 cm^3 of water was poured into a metal can (called a calorimeter) and clamped over the burner. It was noted that 100 cm^3 of water has a mass of 100 g.
2 The starting temperature of the water was measured.
3 The starting mass of the burner and fuel was measured.
4 The burner was lit under the metal can.
5 When the temperature of the water in the can had increased by approximately 30 °C, the burner was extinguished safely.
6 The maximum temperature of the water in the can was measured.
7 The mass of the burner and fuel was measured.

Analysis

To calculate the enthalpy change, the following steps need to be taken:

● Calculate the temperature change of the water.
● Calculate the energy change in the water using the following equation:

heat energy change = mass of water × SHC × temperature change
(J)   (g)   (J/g/°C)   (°C)

$Q = mc\Delta T$

● Convert the heat energy change to kJ by dividing by 1000.
● Calculate the mass change of the burner and fuel, which is equal to the mass of the fuel burned.
● Convert the mass of fuel to the number of moles of fuel using the following equation (see page 26):

$$\text{moles} = \frac{\text{mass (g)}}{M_r}$$

● Calculate the enthalpy change using this equation:

$$\text{enthalpy change} = \frac{-(\text{Heat energy change in kJ})}{\text{moles}}$$

● Remember that the enthalpy of combustion must be negative because it is an exothermic reaction.

> **Exam tip**
>
> The analysis of this experiment assumes that complete combustion has taken place (see page 106) and that all of the energy released from the fuel has been absorbed by the water, i.e. none has been lost to the surroundings. In reality, incomplete combustion usually occurs, and some heat is always lost to the surroundings.

> **Exam tip**
>
> The value of the specific heat capacity for water is 4.2 J/g/°C, but this will be given to you in the exam.

> **Example**
>
> In an experiment, 75 cm^3 of water was heated above a burner containing ethanol. The starting temperature was 22 °C and after 4 minutes the temperature was 45 °C. The mass of the burner and fuel decreased from 145.56 g to 145.10 g. Calculate the enthalpy of combustion.
>
> ● Temperature change = 45 − 22 = 23 °C
> ● $Q = mc\Delta T$
>   = 75 × 4.2 × 23
>   = 7245 J
> ●   = 7.245 kJ
> ● Mass of fuel burned = 145.56 − 145.10 = 0.46 g
> ● Moles = $\dfrac{\text{mass (g)}}{M_r}$
>
> Moles of ethanol burned = $\dfrac{0.46}{46}$ = 0.01
>
> ● Enthalpy change = $\dfrac{-(\text{Heat energy change in kJ})}{\text{moles}}$
>
>   = $\dfrac{-7.245}{0.01}$ = −724.5 kJ/mol

> **Typical mistake**
>
> Students often use the mass of the fuel in the equation for the heat energy change. However, you must use the mass of the water in that equation, which will be equal to its volume in cm^3.

> **Typical mistake**
>
> If you are given the time that the fuel was burned for, ignore it in this calculation as it is irrelevant.

Required practical

Investigate temperature changes accompanying displacement and neutralisation reactions, and salts dissolving in water

Figure 3.5 shows the apparatus used to measure enthalpy changes of reactions that take place in aqueous solutions.

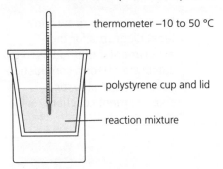

thermometer −10 to 50 °C

polystyrene cup and lid

reaction mixture

Figure 3.5 The apparatus used to measure the enthalpy changes in solutions

Method

For the displacement reaction with zinc and copper sulfate
(i.e. zinc + copper sulfate → copper + zinc sulfate):
1 25 cm³ of copper sulfate solution was poured into a polystyrene cup which was placed inside a beaker so that it did not fall over.
2 The temperature of the solution was recorded.
3 0.5 g of powdered zinc was added to the solution and stirred.
4 The maximum temperature of the mixture was recorded.

Analysis

In this experiment the zinc was the limiting reactant.
● Calculate the temperature change of the solution.
● Calculate the energy change in the water using the following equation:

$$Q = mc\Delta T$$

● Convert the heat energy change to kJ by dividing by 1000.
● Convert the mass of metal to the number of moles using the following equation (see page 26):

$$\text{moles} = \frac{\text{mass (g)}}{A_r}$$

● Calculate the enthalpy change using this equation:

$$\text{enthalpy change} = \frac{-(\text{Heat energy change in kJ})}{\text{moles}}$$

● Remember that the enthalpy change must be negative if the temperature of the solution increased, because it is an exothermic reaction.

Exam tip

A very similar method can be used to calculate temperature changes (and therefore enthalpy changes) for neutralisation reactions and dissolving.

Exam tip

In the exam, you will be told to assume that the specific heat capacity of the solution is the same as for water: 4.2 J/g/°C.

Example

In an experiment, 0.5 g of zinc was added to 25 cm³ of copper sulfate solution. The starting temperature was 24 °C and the maximum temperature reached was 30 °C. Calculate the enthalpy change of this reaction.   →

- Temperature change = $30 - 24 = 6\,°C$
- $Q = mc\Delta T$

$$= 25 \times 4.2 \times 6 = 630\,J$$

$$= 0.63\,kJ$$

$$\text{Moles} = \frac{\text{mass (g)}}{A_r}$$

- Moles of zinc used $= \frac{0.5}{65} = 0.0077$

- Enthalpy change $= \dfrac{-(\text{Heat energy change in kJ})}{\text{moles}}$

$$= \frac{-0.63}{0.0077} = -81.8\text{ kJ/mol}$$

Typical mistake

Remember to be careful about what mass you are using in step 2. For example, in this case the mass (volume) of the copper sulfate solution is used in step 2, not the mass of zinc.

Revision activity

From memory, draw and label diagrams for the experimental method used to calculate ΔH for a combustion reaction, and also for a displacement reaction.

Now test yourself

TESTED

4 Why would it be a good idea to place a lid onto the polystyrene cup containing the chemical reaction when investigating the enthalpy change of an exothermic reaction?

5 $150\,cm^3$ of water was placed in a copper beaker and heated using an alcohol burner. 0.02 moles of propanol were burned, and this caused a temperature increase of $23\,°C$. Calculate the enthalpy of combustion of propanol.

6 A student measured the enthalpy of dissolving of ammonium nitrate. She measured $30\,cm^3$ of water into a polystyrene cup and measured its temperature: $23\,°C$. She then added $2\,g$ of ammonium nitrate (NH_4NO_3) and stirred. The temperature decreased to $11\,°C$. Calculate the enthalpy of dissolving of ammonium nitrate.

Answers on page 127

Enthalpy changes and bonding

Using bond energies to explain exothermic and endothermic reactions

REVISED

The first step in a chemical reaction involves breaking the covalent bonds in the reactant molecules. Breaking bonds requires energy to be absorbed by the reactants, so it is an endothermic process. The second stage in a reaction involves forming the new bonds in the product molecules. Making bonds releases energy, so it is an exothermic process.

This is summarised below:

- Step 1: Breaking old bonds in reactants absorbs energy = endothermic
- Step 2: Making new bonds in products releases energy = exothermic

If less energy is absorbed in step 1 than is released in step 2, the overall reaction is exothermic. If more energy is absorbed in step 1 than is released in step 2, the overall reaction is endothermic.

Typical mistake

Students sometimes refer to energy being needed to make a bond, which is incorrect. Energy is actually released when new bonds are made.

Exam tip

You can remember this energy transfer with the mnemonic 'Breaking and entering'. Burglars **break in** (**breaking** bonds takes energy **in**) and **make out** with their stolen goods (**making** bonds gives energy **out**).

Bond energy calculations

In the exam, you will be given bond energy values, such as those in Table 3.1. These values show the amount of energy needed to break one mole of a specific type of covalent bond.

Table 3.1 Bond energies

Bond	Mean bond energy (kJ/mol)
H–H	436
Cl–Cl	242
H–Cl	431
C–H	413
O=O	498
C=O	805
H–O	464

Example

Calculate the enthalpy change of the following reaction using bond energies given in Table 3.1.

$$H-\underset{\underset{H}{|}}{\overset{\overset{H}{|}}{C}}-H + 2\,O{=}O \longrightarrow O{=}C{=}O + 2\,H\underset{}{\overset{O}{\diagdown}}H$$

Bonds broken in reactants	
Four C–H bonds	4 × 413 = 1652
Two O=O bonds	2 × 498 = 996
Total	2648 kJ

Bonds made in products	
Two C=O bonds	2 × 805 = 1610
Four H–O bonds	4 × 464 = 1856
Total	3466 kJ

- Enthalpy change = (energy in to break bonds) – (energy out when new bonds are made)

$$= 2648 - 3466 = -818\,\text{kJ/mol}$$

- This is therefore an exothermic reaction because the value for ΔH is negative.

Exam tip

Bond energy calculations are easiest to get right if you set out your working in a sensible way like the tables in the example. This layout also makes it easier for the examiner to give you marks for your working out, even if you do not get the final answer right.

Bond energy calculations and energy level diagrams

You can label the bond breaking and bond making steps on an energy level diagram to show why a reaction is exothermic or endothermic. Figure 3.6 is a labelled energy level diagram for the reaction in the example above.

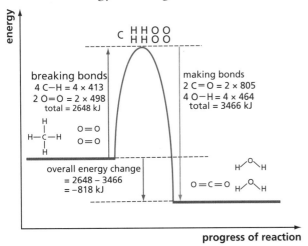

Figure 3.6 A labelled energy level diagram showing how bond energy calculations explain why the reaction is exothermic

Now test yourself

7 In terms of bonds, what is the first process to occur during a chemical reaction? Is this process exothermic or endothermic?
8 What are the units of bond energy values?
9 Hydrogen reacts with chlorine to form hydrogen chloride, according to the following equation:

$$H–H + Cl–Cl \rightarrow 2\ H–Cl$$

Using the bond energies in Table 3.1, calculate the enthalpy change of this reaction.
10 Hydrogen reacts with oxygen to form water, according to the following equation:

$$2\ H–H + O=O \rightarrow 2\ H–O–H$$

Using the bond energies in Table 3.1, calculate the enthalpy change of this reaction.

Answers on pages 127–8

Studying reaction rates

Calculating reaction rates

The rate of a reaction is how fast it happens. The rate of reaction can be calculated using the following equation:

$$\text{reaction rate} = \frac{\text{change in amount of reactant or product}}{\text{time taken}}$$

The units of reaction rate depend on the units of the amount of reactant or product, and the units of time, as shown in Table 3.2.

Table 3.2 How the units of reaction rate can be calculated

Units of amount of product or reactant	Units of time	Units of reaction rate
centimetres cubed, cm³	seconds, s	cm³/s
grams, g	seconds, s	g/s
moles per decimetre cubed, mol/dm³	seconds, s	mol/dm³/s

Example

Calculate the average rate of reaction if 60 cm³ of hydrogen was produced during 80 seconds.

- reaction rate = $\dfrac{\text{change in amount of reactant or product}}{\text{time taken}}$

- reaction rate = $\dfrac{60}{80}$ = 0.75 cm³/s

Measuring the volume of a gaseous product

A reaction that produces a gas can be monitored using the apparatus in Figure 3.7. When the reactants are combined, a stopwatch is started and the volume of gas is recorded at regular intervals (e.g. every 10 seconds).

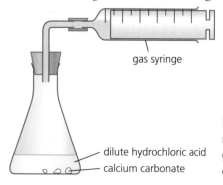

gas syringe

dilute hydrochloric acid
calcium carbonate

Figure 3.7 The apparatus used to monitor a reaction that produces a gas so that the reaction rate can be calculated

Exam tip

If the gas being produced by a reaction is not soluble in water (e.g. hydrogen), the gas can also be collected via a delivery tube in an inverted measuring cylinder full of water which is clamped in a water bath. If the gas is soluble in water, this is not a valid way of measuring the volume of the gas because some of it will dissolve in the water as it bubbles through.

A graph of the results from this experiment is shown in Figure 3.8.

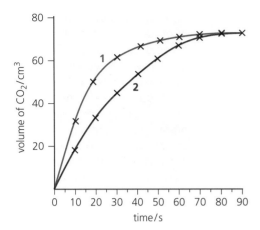

Figure 3.8 A graph of volume of carbon dioxide produced over time during the reaction

The red curve represents a reaction that was faster and the blue curve shows a slower reaction. A faster reaction has a steeper line at the start of the reaction, and it finishes (flattens out) sooner.

Measuring a change in mass

REVISED

If the reaction produces a gas with a sufficiently high relative formula mass (e.g. carbon dioxide) then the decrease in mass as the gas is produced can be easily measured, as shown in Figure 3.9.

If the mass of the flask and contents are plotted against time, the graph looks like Figure 3.10:

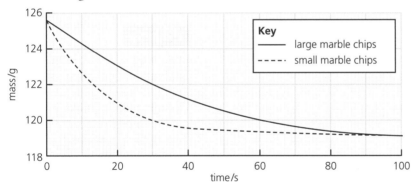

Figure 3.10 A graph of mass of the flask and contents over time during the reaction

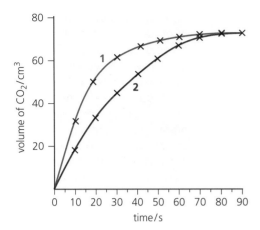

Figure 3.9 The apparatus used to measure the mass decrease in a reaction that produces a gas

If the mass **decrease** is plotted against time, the graph has a similar shape to the one in Figure 3.8.

Reactions that change colour

REVISED

Some aqueous reactions produce a solid product as a precipitate, or a soluble product that has a dark colour. Figure 3.11 shows one method commonly used to judge when a liquid has become opaque because of the colour change. When the reactants are mixed, a stopwatch is started. The observer looks down, through the solution, at the cross drawn on the piece of paper. When the observer can no longer see the cross, the stopwatch is stopped. By using this method, the approximate rate (in s^{-1}) can be calculated using the equation:

$$\text{approximate rate of reaction} = \frac{1}{\text{time taken for cross to disappear}}$$

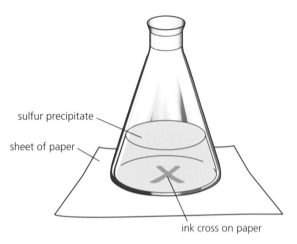

sulfur precipitate

sheet of paper

ink cross on paper

Figure 3.11 The apparatus used to measure the time taken for a solution to become opaque

Calculating the rate of reaction from a graph

A graph of amount of product against time can be used to calculate the reaction rate at any given time. To calculate this, you should draw a **tangent** to the curve at the specific time you are investigating. The gradient of this tangent is then calculated.

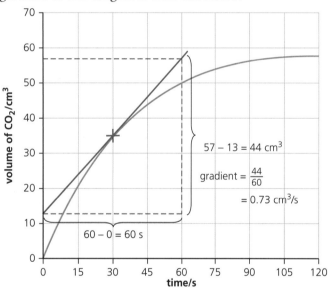

$57 - 13 = 44 \text{ cm}^3$

$\text{gradient} = \dfrac{44}{60}$

$= 0.73 \text{ cm}^3/\text{s}$

$60 - 0 = 60 \text{ s}$

Figure 3.12 A tangent to the curve is drawn in red at $t = 30$ seconds. The gradient of the red tangent is the rate of reaction at 30 seconds

The graph in Figure 3.12 can also be used to calculate the mean rate of reaction. For example, during the first 60 seconds, 50 cm^3 of gas was produced. This means that the average reaction rate over the first 60 seconds is given by:

$$\text{reaction rate} = \frac{\text{change in amount of reactant or product}}{\text{time taken}}$$

$$= \frac{50}{60} = 0.83 \text{ cm}^3/\text{s}$$

> **Revision activity**
>
> Make a revision poster to summarise each of the three different methods described that can be used to investigate the rate of a reaction. For each method, include a diagram, key points to remember from the practical, and a sketch graph of what the results would look like for a fast and a slow reaction.

> **Tangent:** A straight line that touches a curve and follows the curvature of the line at that point but extends away from the curve and does not cross it.

> **Exam tip**
>
> When using this method to calculate a gradient, you should make your triangle as large as possible to get an accurate value. It also makes the maths easier if you extend the tangent to touch the *x*- or *y*-axis.

Now test yourself

11 By interpreting the state symbols in the reaction below, suggest a method that could be used to measure the rate of reaction.

$$Na_2CO_3(s) + 2HCl(aq) \rightarrow 2NaCl(aq) + H_2O(l) + CO_2(g)$$

12 Look at the equation below for the reaction between hydrochloric acid and sodium thiosulfate. The sulfur dioxide is very soluble and dissolves in the solution, so no bubbles are seen. Suggest a method that could be used to measure the rate of reaction.

$$Na_2S_2O_3(aq) + 2HCl(aq) \rightarrow 2NaCl(aq) + S(s) + SO_2(g) + H_2O(l)$$

13 Calculate the average rate of reaction during the first 20 seconds of the reaction shown in Figure 3.13. Include your units.

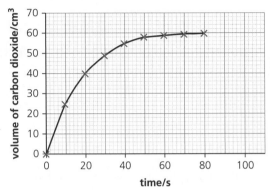

Figure 3.13

14 Using the same graph, calculate the rate of reaction at 30 seconds by drawing a tangent to the curve. Include your units.

Answers on page 128

Concentration, pressure and reaction rates

Collision theory in solutions

Chemists use collision theory to explain why reaction rates are affected by certain factors. Collision theory states that:

- For a reaction to occur between two particles, they need to collide with a minimum amount of energy, which is called the **activation energy**.
- Factors that increase the frequency of collisions increase the reaction rate.
- Factors that increase the proportion of particles that have more energy than the activation energy increase the reaction rate.

Increasing the concentration of a dissolved reactant (e.g. an acid) will increase the rate of a reaction because:

- A higher concentration means that there is a greater number of reactant particles per unit volume of solution (the particles are more crowded).
- Therefore, successful collisions between particles occur more frequently.

> **Activation energy:** The minimum energy needed by particles to react together when they collide. If they have an amount of energy less than the activation energy, they simply bounce off each other.

> **Exam tip**
>
> The explanation in these two bullet points is often required in exam questions. You are therefore strongly advised to learn it carefully. Typically, these bullet points are worth one mark each.

Collision theory in gases

Collision theory is also used to explain why reactions in gases occur more quickly at higher pressures:

- A higher pressure means that there is a greater number of reactant particles per unit volume of gas (the particles are more crowded).
- Therefore, successful collisions between particles occur more frequently.

Exam tip

You will notice that this explanation is almost identical to the bullet points which explain the effect of increasing the concentration of a solution.

Required practical

Investigate the effect of changing the concentration of hydrochloric acid on the rate of reaction between marble chips and dilute hydrochloric acid

In this experiment, marble chips (calcium carbonate, $CaCO_3$) were reacted with hydrochloric acid, according to the following equation:

$$CaCO_3(s) + 2HCl(aq) \rightarrow CaCl_2(aq) + H_2O(l) + CO_2(g)$$

Method

1 The apparatus was set up as shown in Figure 3.9 on page 91.
2 At the moment that the reactants were mixed in the conical flask and the bung was replaced, a stopwatch was started.
3 The volume of gas in the syringe was recorded every 10 seconds until the reaction had finished.
4 The experiment was repeated with a different concentration of acid.

Analysis

- A graph of volume of gas (y-axis) was plotted against time (x-axis).
- The different data sets (from the different concentrations of acid) were plotted as separate curves.
- The rate of reaction at the start of each reaction can be calculated by working out the gradient of a tangent drawn to the curve at $t = 0$ seconds.
- The reaction with the more concentrated acid had a steeper curve, which tells us that the reaction was faster.

Now test yourself

15 What is meant by the term 'activation energy'?
16 A student reacted $20 \, cm^3$ of $0.1 \, mol/dm^3$ sulfuric acid with a 2 cm long piece of magnesium ribbon. The reaction took 43 seconds to finish. The student then repeated the experiment with $20 \, cm^3$ of $1 \, mol/dm^3$ sulfuric acid. Suggest the time taken for the second reaction to finish.
17 Explain your answer to question 16, using ideas about collisions between particles.

Answers on page 128

Temperature, catalysts and reaction rates

The effect of temperature on reaction rate

The effect of temperature on the rate of a chemical reaction can be investigated using any of the methods described on pages 90–92.

Increasing the temperature always increases the rate of reaction. This can be explained by collision theory. When a reaction occurs at a higher temperature:

● The particles have more kinetic energy, so they move faster.
● This means that successful collisions occur more frequently.
● It also means that a greater proportion of the collisions that take place will be successful, because a greater proportion of the particles will have more energy than the activation energy required.

> **Exam tip**
>
> The concept of activation energy can be explained with an analogy. Imagine a crowd of people (particles), each hoping to get into a concert (to react). Each person has a different amount of money in their pocket. This is the amount of energy that each particle has. The entry fee is fixed – this is the activation energy. Some people have enough money to get in, others do not. But if you give everyone a little more money (energy), a greater proportion of the crowd can get into the concert.

The effect of catalysts on reaction rate

REVISED

A **catalyst** is a chemical that speeds up a reaction without being used up or becoming chemically changed at the end of the reaction. Catalysts are usually specific, which means that they may only speed up one chemical reaction. Catalysts are often transition metals, or compounds of transition metals.

Catalysts work by providing an **alternative pathway** for the reaction – one which requires a lower activation energy. This means that a greater proportion of the particles have enough energy to react when they collide.

> **Catalyst:** A substance that speeds up a reaction without being used up or becoming chemically changed by the end of the reaction.
>
> **Alternative pathway:** A different chemical route for a reaction, perhaps involving the formation of an intermediate chemical which then changes into a product.

> **Typical mistake**
>
> In exams, you could be asked to describe *what* a catalyst is or to explain *how* a catalyst works. These are very different questions and so needs different answers:
> **What:** A catalyst is a chemical that speeds up a reaction without being used up.
> **How:** A catalyst does this by providing an alternative reaction pathway with a lower activation energy.

> **Typical mistake**
>
> Students often say 'the particles get more activation energy' which does not make sense. The activation energy is the amount of energy they **need**, not the amount of energy they **have**. Instead, you should write that particles gain more **energy**, so a greater proportion of them have more energy than the activation energy required.

> **Exam tip**
>
> The effect of a catalyst can be explained with the same concert analogy used for the effect of temperature. This time, instead of giving everyone in the crowd some more money (energy) by increasing the temperature, the entry fee for the concert (the activation energy) is reduced.

The effect of a catalyst can be seen on the energy level diagrams in Figure 3.14 on the following page.

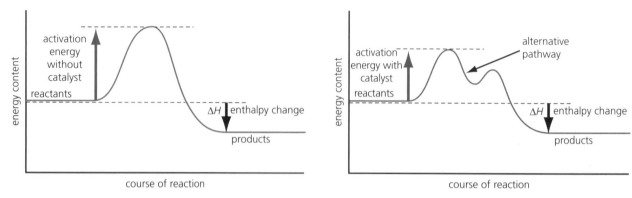

Figure 3.14 The reaction profiles for an exothermic reaction without a catalyst (left) and with a catalyst (right). Note that the enthalpy change is the same in both diagrams, but the activation energy is reduced when a catalyst is used

---

Required practical

Investigate the effect of different solids on the catalytic decomposition of hydrogen peroxide solution

Hydrogen peroxide, H_2O_2, decomposes slowly at room temperature, according to the following equation:

$$2H_2O_2(aq) \rightarrow 2H_2O(l) + O_2(g)$$

Because oxygen gas is produced in this reaction, bubbles will be seen if the reaction occurs quickly. Different solids can be investigated to see which speeds up the reaction the most.

Method

1 Five test tubes were placed in a rack and $10\,cm^3$ of hydrogen peroxide solution was added to each test tube.
2 0.5 g of one of the solids was added to the first test tube. Any observations were recorded. At the end of the reaction, the mixture was filtered and the solid was rinsed and dried.
3 Step 2 was then repeated using 0.5 g of a different solid, added to the second test tube, and so on with other potential catalysts.

Analysis

The results were as follows:

Potential catalyst	Observations	Mass of solid after experiment (g)
Copper(II) oxide	Some small bubbles form. Solid does not dissolve.	0.5
Potassium iodide	Vigorous bubbles. Solution turns brown. Felt hot.	0.1
Sodium chloride	No bubbles. Solid dissolves.	0.0
Manganese(IV) oxide	Vigorous bubbles. Test tube felt hot. Solid does not dissolve.	0.5
Magnesium oxide	No bubbles. Solid does not dissolve.	0.5

This meant:
- The best catalyst from the five solids tested was manganese(IV) oxide because it sped up the reaction, but it was not used up or chemically changed at the end of the reaction.
- Copper(II) oxide also acted as a catalyst, but it was not as effective as the manganese(IV) oxide.
- The potassium iodide caused oxygen gas to be produced very quickly via a different chemical reaction with different products, but it did not act as a catalyst because it was used up during the reaction.

Now test yourself

18 Use ideas about collisions between particles to explain why the chemical reactions that cause food to decay are slower in a refrigerator, compared to if the food was stored at room temperature.

19 Iron is used as a catalyst in the manufacture of ammonia from nitrogen and hydrogen. Explain why and how iron acts as a catalyst in this reaction.

Answers on page 128

Particle size and reaction rates

Particle size and collision theory

REVISED ☐

When one of the reactants is a solid, the size of the particles affects the rate of reaction. The smaller the particles of a solid, the faster the reaction. Powders react very quickly indeed, sometimes explosively.

As the particles of a solid get smaller, their surface area increases. This means that successful collisions between the particles in a solution and the particles of the solid are more frequent when smaller particles are used.

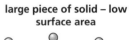

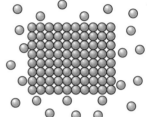

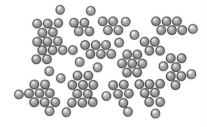

Figure 3.15 The effect of particle size on the rate of reaction

Required practical

Investigate the effect of changing the surface area of marble chips on the rate of reaction between marble chips and dilute hydrochloric acid

In this experiment, calcium carbonate ($CaCO_3$) reacts with hydrochloric acid, according to the following equation:

$$CaCO_3(s) + 2HCl(aq) \rightarrow CaCl_2(aq) + H_2O(l) + CO_2(g)$$

The mass of calcium carbonate used was the same in both experiments below. The calcium carbonate was in the form of large marble chips during the first experiment, and much smaller marble chips in the second experiment. Remember that **smaller** marble chips have a **larger** surface area than larger marble chips.

Method

1 The apparatus was set up as shown in Figure 3.9 on page 91.
2 At the moment that the reactants were mixed in the conical flask on the top pan balance, a stopwatch was started.
3 The mass of the conical flask and contents was recorded every 10 seconds until the reaction finished.
4 The experiment was then repeated with small marble chips. →

Analysis

The results were analysed as follows:

- A graph of mass of the flask and contents (y-axis) was plotted against time (x-axis).
- The different data sets (from the different sized marble chips) were plotted as separate curves.
- The rate of reaction at the start of each reaction can be calculated by working out the gradient of a tangent drawn to the curve at $t = 0$ seconds.
- The reaction using the smaller marble chips gave a line with a steeper curve, which tells us that the reaction was faster.

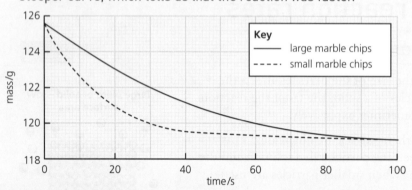

Key
— large marble chips
---- small marble chips

Figure 3.16 A graph to show the effect of particle size on the rate of reaction

Revision activity

Make five flashcards for the different ways to speed up a reaction. On the reverse, you should have two or three bullet points of explanation per method.

Now test yourself

TESTED

20 State which has the larger surface area: lumps or powder.
21 Suggest a reason why catalysts are often used in the form of powders or gauzes to speed up chemical reactions.
22 Refer to Figure 3.16. Calculate the rate of reaction at the start of the experiment for both the large and the small marble chips.

Answers on page 128

Reversible reactions

Many chemical reactions are irreversible, which means that they take place in one direction but cannot occur in the reverse direction. Examples of irreversible reactions include combustion reactions:

methane + oxygen → carbon dioxide + water

$$CH_4 + 2O_2 \rightarrow CO_2 + 2H_2O$$

Most physical changes are easily reversible, by changing the temperature or pressure. When a reaction or process is reversible, we use a double-headed arrow:

water ⇌ steam

$$H_2O(l) \rightleftharpoons H_2O(g)$$

Some chemical changes are also **reversible**, so they can be described using the same symbol ($\rightleftharpoons$).

Reversible reaction: A chemical change which can occur both ways: the products can turn back into the reactants, and vice versa.

The reversible hydration of copper sulfate

For example, anhydrous copper sulfate is a white solid, but when water is added to it, it turns into a blue solid. This chemical reaction is highly **exothermic**. If the blue solid is then heated, an **endothermic** reaction occurs, and the water is lost from the solid, reversing the original reaction and turning the solid back to white:

anhydrous copper(II) sulfate + water $\rightleftharpoons$ hydrated copper(II) sulfate

$$CuSO_4(s) \qquad\quad + H_2O_4(l) \rightleftharpoons \qquad CuSO_4.5H_2 O(s)$$
$$\text{white} \qquad\qquad\qquad\qquad\qquad \text{blue}$$

> **Exothermic:** A process which releases energy (usually as heat) from the chemicals into the surroundings.
>
> **Endothermic:** A process which absorbs energy (usually as heat) from the surroundings into the chemicals.

The above forward reaction is often used as a test for the presence of water (see page 81).

The reversible decomposition of ammonium chloride

Ammonium chloride (NH_4Cl) is a white solid. When it is heated it decomposes into two gases: hydrogen chloride and ammonia. If these gases are allowed to cool, they react together again to re-form the ammonium chloride:

ammonium chloride $\rightleftharpoons$ ammonia + hydrogen chloride

$$NH_4Cl(s) \qquad \rightleftharpoons NH_3(g) + \qquad HCl(g)$$

Now test yourself

23 State the meaning of the term 'reversible reaction'.
24 Write the symbol used to describe a reversible reaction in an equation.
25 In the manufacture of methanol (CH_3OH), carbon monoxide (CO) and hydrogen are reacted together in a reversible reaction in which all the products and reactants are gases. Methanol is the only product. Write a word equation for this reaction.
26 Write a balanced symbol equation with state symbols for the reaction described in question 25.

Answers on page 128

Dynamic equilibria

Explaining dynamic equilibrium

If a reversible reaction occurs within a sealed container, a **dynamic equilibrium** is established. A dynamic equilibrium is when the amounts (or concentrations) of the products and reactants remain constant over time because the forward and reverse reactions are occurring at the same rate.

> **Dynamic equilibrium:** When the forward and reverse reactions occur at the same rate, and the concentrations of the products and reactants remain constant over time.

Typical mistake

Often, students think that in a dynamic equilibrium, the amount of products is the same as the amount of reactants. This is not true.

At dynamic equilibrium, if there is a lot of product present in the mixture, we say that the position of equilibrium lies to the right. If there is a lot of reactant in the equilibrium mixture, we say that the position of equilibrium lies to the left.

The effect of a catalyst

Adding a catalyst to a reversible reaction in a sealed container will speed up the forward and reverse reactions equally. This means that the time taken for the reaction to reach dynamic equilibrium is faster, but the position of the equilibrium is not affected.

Typical mistake

At dynamic equilibrium, you should not say that the reaction has finished. In fact, both the forward and reverse reactions are still happening.

Revision activity

Make definition flashcards for the key terms in this topic, with the term on one side and the definition on the other.

Factors that affect the position of equilibrium in gaseous reactions

REVISED

The position of a dynamic equilibrium can be affected by changing a number of factors.

Changing pressure

Increasing the pressure within a sealed container will cause the position of equilibrium to move towards the side of the equation with the **fewest** moles of gas. Decreasing the pressure will push the reaction in the direction of the greater number of moles of gas.

Example

Predict the effect of increasing the pressure on the following reaction:

$$CO(g) + 2H_2(g) \rightleftharpoons CH_3OH(g)$$

- Increasing the pressure will cause the position of equilibrium to move towards the side with the fewest moles of gas.
- The right-hand side has fewer moles (because there is one mole on the right and three moles on the left).
- Therefore, the amount of methanol (CH_3OH) will increase.

Changing temperature

Increasing the temperature within a sealed container will cause the position of equilibrium to move in the endothermic direction. Decreasing the temperature will push the position of equilibrium towards the exothermic direction.

Exam tip

It is helpful to think of the chemical reaction pushing back against the change that you are trying to impose on it. If you try to increase the pressure, it will try to decrease it by moving toward the side with the fewest moles of gas. If you increase the temperature, it will try to decrease it by moving in the endothermic direction.

Example

Predict the effect of decreasing the temperature on the following reaction, where the forward reaction is exothermic:

$$CO(g) + 2H_2(g) \rightleftharpoons CH_3OH(g)$$

- Decreasing the temperature will cause the position of equilibrium to move towards the exothermic direction.
- This direction is the forward reaction.
- Therefore, the amount of methanol (CH_3OH) will increase.

Case study: making ammonia

Ammonia (NH_3) is made from nitrogen and hydrogen, as described by the following equation. The forward reaction is exothermic:

$$N_2(g) + 3H_2(g) \rightleftharpoons 2NH_3(g)$$

The effects of pressure and temperature on the reaction mixture at equilibrium are shown by the graph in Figure 3.17.

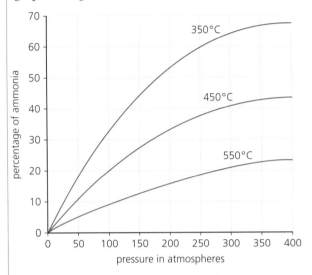

Figure 3.17 A graph to show the effects of pressure and temperature on the amount of ammonia in the reaction mixture at equilibrium

The effect of pressure

The graph in Figure 3.17 shows that at any given temperature, increasing the pressure produces more ammonia.

Increasing the pressure will favour the side with the fewest moles of gas. There are two moles of gas on the right and four on the left, therefore increasing the pressure makes more ammonia.

The effect of temperature

The graph in Figure 3.17 shows that at any given pressure on the x-axis, increasing the temperature produces less ammonia, as you move from one red curve to the one below it (which represents a higher temperature).

Increasing the temperature will favour the endothermic reaction, which is the reverse reaction. This would make more nitrogen and hydrogen and produce less ammonia. Since ammonia is the desired product, a relatively low temperature is used, because this favours the exothermic direction (to the right). However, if the temperature is too low, the reaction rate will be too slow, so a compromise needs to be made in the manufacturing process.

Adding a catalyst

An iron catalyst is used in the manufacture of ammonia. Remember that this speeds up the time taken to reach equilibrium, but does not affect the amount of ammonia in the mixture at equilibrium.

Now test yourself

TESTED ☐

27 State the meaning of the term 'dynamic equilibrium'.
28 Hydrogen and iodine gases are mixed and sealed in a glass container. These gases react according to the following reaction and a dynamic equilibrium is established:

$$H_2(g) + I_2(g) \rightarrow 2HI(g)$$

State and explain the effect of increasing the pressure on this reaction.
29 Sulfur dioxide reacts with oxygen to make sulfur trioxide. The forward reaction is exothermic:

$$2SO_2(g) + O_2(g) \rightarrow 2SO_3(g)$$

State and explain whether a low or high temperature should be used to produce the maximum amount of sulfur trioxide.

Answers on page 128

Exam tip

You do not need to know the details of the process described here to make ammonia, so you should not learn the conditions for this reaction. This case study is simply used to illustrate the points described above. In your exam, you are likely to have to apply your understanding to a reaction that you have not studied before.

Summary

- Exothermic reactions release energy. Endothermic reactions absorb energy. Energy level diagrams (energy profiles) can be used to describe enthalpy changes.
- Enthalpy changes can be calculated by measuring temperature changes in water.
- A reaction is exothermic if the energy absorbed to break bonds in the reactants is less than the energy released when new bonds are made in the products.
- Rates of reaction can be calculated by measuring the amount of product made during a specific time.
- The volume of a gaseous product can be measured using a gas syringe. The mass decrease caused by carbon dioxide escaping from the reaction is also easy to measure. The time taken for a colour change to occur in a reaction is also easy to measure by observing transparency changes.
- Increasing the concentration of a dissolved reactant increases the rate of reaction. This is because there are more reactant particles per unit volume, so successful collisions are more frequent.
- Increasing the pressure of gaseous reactants increases the rate of reaction. This is because there are more reactant particles per unit volume, so successful collisions are more frequent.

- Increasing the temperature increases the rate of reaction. This is because the particles have more energy, so successful collisions are more frequent. Also, a greater proportion of particles have an amount of energy which is equal to or greater than the activation energy, so a greater proportion of the collisions will be successful.
- A catalyst speeds up a reaction but is not used up or chemically changed at the end of the reaction. Catalysts work by providing an alternative pathway with a lower activation energy.
- Powders react faster than lumps. This is because smaller particles have a larger surface area, so successful collisions are more frequent.
- Some chemical reactions are reversible ($\rightleftharpoons$). A reversible reaction in a sealed container will result in a dynamic equilibrium, where the forward and reverse rates are equal and the concentrations and amounts of products and reactants remain constant over time.
- A catalyst speeds up the time taken to reach equilibrium, but does not affect the position of equilibrium.
- Increasing the pressure moves the position of equilibrium in the direction of the side with the fewest moles of gas.
- Increasing the temperature moves the position of equilibrium in the endothermic direction.

Exam practice

1 Ethanol is a liquid fuel with the molecular formula C_2H_5OH.
  a The enthalpy of combustion of ethanol can be calculated by adding ethanol to a burner and using it to heat water in a suitable container. Describe how you could measure the enthalpy of combustion of ethanol in this way. You should include the measurements you would take in the experiment in your answer. You may use a labelled diagram. [6]
  b The enthalpy of combustion of ethanol can also be calculated using a bond energy calculation. The balanced symbol equation is given below, and the bond enthalpies are provided in the table.

$$
\begin{array}{c}
\text{H} \quad \text{H} \\
| \quad | \\
\text{H}-\text{C}-\text{C}-\text{O}-\text{H} + 3\,\text{O}=\text{O} \longrightarrow 2\,\text{O}=\text{C}=\text{O} + 3\,\text{H}-\text{O}-\text{H} \\
| \quad | \\
\text{H} \quad \text{H}
\end{array}
$$

Bond	Mean bond energy (kJ/mol)
C–H	413
O=O	498
C=O	805
H–O	464
C–O	358
C–C	347

Calculate the enthalpy of combustion of ethanol. [4]

c The value for the enthalpy of combustion obtained from the bond enthalpy calculation is more exothermic than the results of the experiment. Suggest **two** reasons why this might be the case. [2]

2 A student uses the following method in an experiment to measure the enthalpy change of neutralisation:
   ● 25 cm³ of dilute hydrochloric acid was measured into a polystyrene cup which was placed in a glass beaker.
   ● 25 cm³ of dilute sodium hydroxide was measured into a separate cup.
   ● The concentrations of the two solutions were 1 mol/dm³.
   ● The temperature of both solutions was 22 °C.
   ● The two solutions were mixed and stirred. The temperature increased to 31 °C.
   a State whether this reaction is exothermic or endothermic. [1]
   b Assume that the solution has a density of 1 g/cm³ and a specific heat capacity of 4.2 J/g/°C. Calculate the energy change in joules during the reaction. [3]
   c During the reaction, 0.025 moles of acid were neutralised. Calculate the enthalpy change of neutralisation. [2]
   d Draw and label an energy level diagram for this reaction. [4]

3 The reaction between hydrochloric acid and sodium thiosulfate can be used to investigate the effect of changing the concentration of sodium thiosulfate on the rate of reaction.
   $$HCl + Na_2S_2O_3 \rightarrow NaCl + S + SO_2 + H_2O$$
   The reaction mixture turns from colourless to opaque pale yellow due to the insoluble sulfur particles. The reaction takes place in a conical flask which is placed on a piece of white paper that has a cross drawn on it as in Figure 3.11 on page 92. The observer looks down through the solution and stops the stopclock when the cross is no longer visible through the reaction mixture.
   a Balance the equation for this reaction.
      ___HCl + ___$Na_2S_2O_3$ → ___$NaCl$ + ___S + ___SO_2 + ___H_2O [1]
   b State **three** variables that must be controlled in this reaction if the experiment is to be valid. [3]
   c The results obtained from this experiment are shown below. Complete a copy of the table to show the rate of reaction. [2]

Concentration of $Na_2S_2O_3$ (mol/dm³)	Time taken until cross is no longer visible (s)	Rate of reaction (s^{-1})
0.25	58	
0.20	72	
0.15	98	
0.10	151	
0.05	320	

   d Plot the values of rate of reaction on a graph. Plot rate on the *y*-axis against concentration of sodium thiosulfate on the *x*-axis. Draw a straight line of best fit for the points you have plotted. [3]
   e Explain how increasing the concentration of sodium thiosulfate affects the rate of the reaction. [3]

4 Nitrogen dioxide (NO_2) is a dark brown gas. Two molecules of nitrogen dioxide can react to make the colourless gas dinitrogen tetroxide (N_2O_4). The reaction is shown below. The forward reaction is exothermic.
   $$2NO_2 \rightleftharpoons N_2O_4$$
   a What does the $\rightleftharpoons$ symbol mean? [1]
   b When a mixture of NO_2 and N_2O_4 is placed inside a sealed gas syringe, a dynamic equilibrium is established. State what is meant by the term 'dynamic equilibrium'. [2]
   c The mixture in the gas syringe has a pale brown gas due to the NO_2 which is present. When the plunger is pressed to compress the mixture, the colour fades to a very pale brown. Explain why this colour change occurs. [2]
   d State and explain whether a higher or a lower temperature would cause a mixture of NO_2 and N_2O_4 to go darker. [3]

Answers and quick quizzes online

ONLINE ☐

4 Organic chemistry

Crude oil

What is crude oil made from?

REVISED

Crude oil is a fossil fuel which has formed from the remains of tiny living organisms over millions of years. Crude oil is a mixture of **hydrocarbon** compounds.

> **Hydrocarbon:** A compound that is made from hydrogen and carbon atoms **only**.

> **Exam tip**
>
> You might be asked to identify which compound (from a list) is a hydrocarbon. To answer this question you should look for a compound which has **only** C and H atoms in it.

Fractional distillation

REVISED

The hydrocarbons in crude oil have very different boiling points, from below −40 °C to above 300 °C. This means that they can be separated by **fractional distillation**. The gases and liquids produced from the fractional distillation of crude oil are called **fractions**. Each fraction is still a mixture of hydrocarbon compounds, but those compounds have very similar properties (e.g. boiling point, viscosity, flammability) so they can be used for a specific purpose.

> **Fractional distillation:** A technique used to separate a mixture of liquids that all have different boiling points.
>
> **Fraction:** A mixture of hydrocarbons which have similar boiling points and are obtained from fractional distillation.

The industrial fractional distillation of crude oil is shown in Figure 4.1. The steps in this process are as follows:

1 Crude oil is heated until it turns into a mixture of gases. It is then fed into the fractionating tower near to the bottom.
2 As the gases rise up the tower, they cool down.
3 Compounds that cool to a temperature equal to their boiling point condense into a liquid and leave the tower.
4 Any compounds that reach the top (coolest part) of the tower (and so are still gases) are piped off as refinery gases.

> **Typical mistake**
>
> Students sometimes think that the higher up the fractionating tower the hotter it will be, because they remember that hot gases and liquids rise up due to a lower density. What they forget is that the crude oil is heated **before** entering the tower. As it rises, it cools, just like the way that steam from a kettle cools as it rises. The steam is hottest low down, near the source of heat (the kettle).

The compounds that have the **longest molecules** have the **strongest intermolecular forces**, so they have the **highest boiling points** and will condense lower down the tower. This is because more energy is needed to keep their molecules separated as a gas.

> **Exam tip**
>
> You can make it easier to remember the link between molecule size and boiling point by repeating the following words: '**longer – stronger – higher**'.

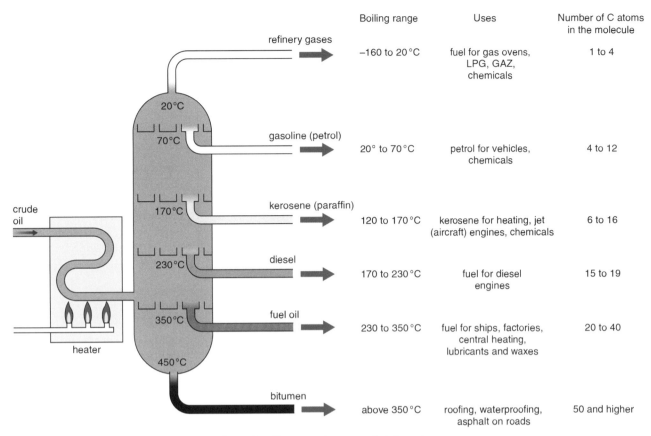

	Boiling range	Uses	Number of C atoms in the molecule
refinery gases	−160 to 20 °C	fuel for gas ovens, LPG, GAZ, chemicals	1 to 4
gasoline (petrol)	20° to 70 °C	petrol for vehicles, chemicals	4 to 12
kerosene (paraffin)	120 to 170 °C	kerosene for heating, jet (aircraft) engines, chemicals	6 to 16
diesel	170 to 230 °C	fuel for diesel engines	15 to 19
fuel oil	230 to 350 °C	fuel for ships, factories, central heating, lubricants and waxes	20 to 40
bitumen	above 350 °C	roofing, waterproofing, asphalt on roads	50 and higher

Figure 4.1 The industrial fraction of crude oil to produce useful fractions

Revision activity

Copy the names of the different fractions, their boiling points, uses, and typical number of carbon atoms (from Figure 4.1) onto different cards. Shuffle the pack and then try to match them up. See if you can also put the fractions into the correct order.

There are clear trends in the properties of the fractions. As the fractions get longer:

- They get less volatile (their boiling points increase).
- They get darker in colour.
- They get more viscous.

Now test yourself

TESTED ☐

1 Which of the following compounds are hydrocarbons:
CH_4, CH_3COOH, C_6H_6, $C_2H_5NH_2$, $C_6H_{12}O_6$?
2 State the name of the process used to separate crude oil into useful fractions.
3 Name the fraction which is used as fuel for aeroplanes.
4 State the link between the length of the molecules and the viscosity of the fraction.

Answers on page 129

> **Volatile:** A liquid that has a low boiling point, so it evaporates easily.
>
> **Viscous:** A liquid that pours very slowly. Water has a low viscosity and pours quickly. Syrup has a high viscosity and pours slowly.

How pure is our air?

Complete and incomplete combustion

REVISED

A **fuel** is a substance that is burned to release energy. The products of combustion depend on the fuel that is being burned and the amount of oxygen available. For example, where plenty of oxygen is available, combustion may occur as follows:

carbon + oxygen → carbon dioxide

$$C + O_2 \rightarrow CO_2$$

hydrogen + oxygen → water

$$2H_2 + O_2 \rightarrow 2H_2O$$

Because a hydrocarbon fuel contains both carbon and oxygen, it will produce carbon dioxide and water, as long as there is a good supply of oxygen. This is called **complete combustion** and the maximum amount of energy is released.

If there is insufficient oxygen for complete combustion, **incomplete combustion** occurs. During incomplete combustion, less energy is released than in complete combustion. The products of incomplete combustion of a hydrocarbon fuel are water, plus carbon monoxide or carbon (or a mixture of both). The carbon produced is sometimes described as **particulate carbon**, or soot.

> **Example**
>
> Write a balanced symbol equation to show the incomplete combustion of propane (C_3H_8) to produce water and carbon monoxide.
>
> - $C_3H_8 + O_2 \rightarrow CO + H_2O$
> - When balanced, this is:
>
> $$2C_3H_8 + 7O_2 \rightarrow 6CO + 8H_2O$$

> **Fuel:** A substance that is burned in order to release energy.
>
> **Complete combustion:** When there is a good supply of oxygen, complete combustion of a hydrocarbon fuel produces carbon dioxide and water.

> **Example**
>
> Write a word equation to show the complete combustion of the hydrocarbon fuel butane.
>
> - Butane + oxygen → carbon dioxide + water

> **Incomplete combustion:** When there is a poor supply of oxygen, incomplete combustion of a hydrocarbon fuel occurs, which produces water, plus carbon monoxide or carbon.

Investigating the products of combustion

The apparatus shown in Figure 4.2 can be used to investigate the products of combustion of candle wax. Particulate carbon (soot) forms on the thistle funnel, showing that incomplete combustion is occurring. Water condenses in the U-tube, and if this is added to anhydrous copper sulfate, the copper sulfate turns from white to blue. Carbon dioxide is detected by the limewater, which turns cloudy.

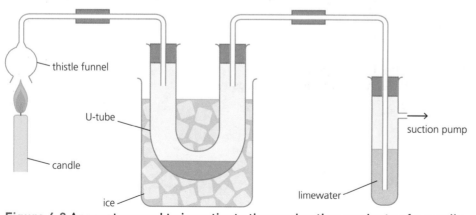

Figure 4.2 Apparatus used to investigate the combustion products of a candle

Disadvantages of incomplete combustion

There are three disadvantages of incomplete combustion:

1. Less energy is released.
2. Carbon monoxide is released. Carbon monoxide is poisonous because if it is inhaled, it is absorbed by red blood cells, preventing them from transporting oxygen as effectively.
3. Carbon is produced in the form of soot. Soot is dirty and can block flues and chimneys.

Revision activity

Draw up a summary table to compare complete and incomplete combustion. Include in your table: what causes each type of combustion, how much energy is released, as well as the chemical products.

Acid rain caused by sulfur dioxide

REVISED

Some fossil fuels contain sulfur impurities. When the fuel is burned, the sulfur atoms are oxidised and produce sulfur dioxide:

sulfur + oxygen → sulfur dioxide

$$S + O_2 \rightarrow SO_2$$

Sulfur dioxide reacts with more oxygen and water in the atmosphere to produce sulfuric acid:

sulfur dioxide + oxygen + water → sulfuric acid

$$2SO_2 + O_2 + 2H_2O \rightarrow 2H_2SO_4$$

Sulfuric acid is one of the major causes of **acid rain**. Acid rain damages plants and animals, especially those in ponds, lakes and rivers. It also damages statues and buildings.

Typical mistake

Students sometimes think that sulfur is added to fuels deliberately as an additive. This is not true. Scientists try very hard to remove sulfur from fuels before they are sold.

Acid rain: Rainwater that has a pH significantly lower than 7.

Acid rain caused by nitrogen oxides

REVISED

Nitrogen makes up 78% of air, but it does not normally react with oxygen because it is unreactive. However, inside a car engine, the temperature is so high that atmospheric nitrogen and oxygen react together to produce nitrogen oxides.

The nitrogen oxides which are produced are collectively known as NO_x. NO_x reacts with water in the atmosphere to produce nitric acid (HNO_3), which contributes to acid rain.

Typical mistake

Students often think that nitrogen is present in the fuel being burned. This is not true. In these reactions, nitrogen has come from the air, not the fuel.

Now test yourself

TESTED

5. Write a word equation for the complete combustion of the hydrocarbon fuel octane.
6. Write a balanced symbol equation for the complete combustion of propane, C_3H_8.
7. Describe three disadvantages of incomplete combustion compared with complete combustion.
8. Suggest why removing sulfur from diesel fuel is a legal requirement in the UK.

Answers on page 129

Revision activity

Construct a large table which lists the following pollutants in the first column: carbon dioxide, carbon monoxide, soot/particulate carbon, sulfur dioxide, nitrogen oxides. The headers for the remaining columns should be: source, and harmful effects. Fill this in to help your revision.

Cracking – more petrol from crude oil

Matching supply with demand

Fractional distillation is used to obtain useful fractions from crude oil. Most of these fractions are used as fuels. Fuels made from shorter hydrocarbons tend to make more convenient fuels because they are easier to burn and more likely to undergo complete combustion, releasing the maximum amount of energy. This means that the demand for shorter hydrocarbons is actually greater than the amount produced after fractional distillation. For the longer hydrocarbons, the amount available (the supply) is greater than the demand. This means there is a surplus of the long hydrocarbon fractions.

In other words:

- Short hydrocarbons: supply < demand
- Long hydrocarbons: supply > demand

To solve this problem, long hydrocarbons are decomposed (broken down) into shorter fractions, using an industrial process called **cracking**.

> **Cracking:** Converting long, less useful hydrocarbons into smaller, more useful hydrocarbons.

The conditions for cracking

The chemical reactions in cracking are **thermal decomposition** reactions, and they take place at high temperatures (600–700 °C) in the presence of a catalyst (alumina or silica). The products are shorter hydrocarbons that can be grouped into two 'families' called **alkanes** and **alkenes**. You will find more detail on alkanes and alkenes in the following sections.

> **Thermal decomposition:** Breaking down one compound into two or more substances using heat.
>
> **Alkane:** A type of hydrocarbon in which the carbon atoms are all bonded by single C–C bonds.
>
> **Alkene:** A type of hydrocarbon which contains at least one C=C double bond.

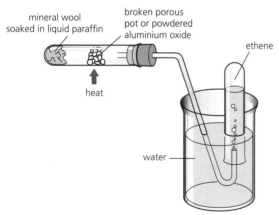

labels: mineral wool soaked in liquid paraffin; broken porous pot or powdered aluminium oxide; ethene; heat; water

Figure 4.3 Apparatus used to demonstrate cracking in the school laboratory

Equations for cracking

Cracking can be summarised with the following general word equation:

long alkane → short alkane + short alkene

The short alkanes are particularly useful in fuels like petrol. The short alkenes are used as reactive compounds that are the starting points for many useful chemical reactions, such as addition polymerisation (see page 118).

Cracking can be undertaken with a variety of long alkanes as the starting point. Because the chain can be broken in a variety of places during the reaction, we can produce a variety of products. One example reaction is shown in Figure 4.4.

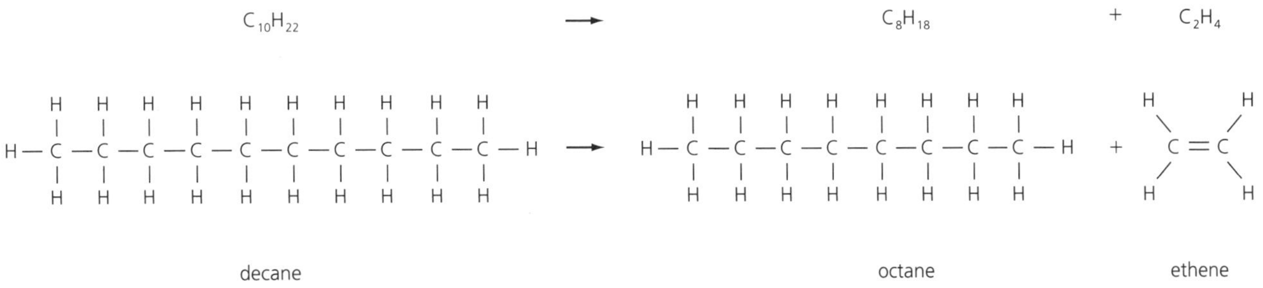

$$C_{10}H_{22} \longrightarrow C_8H_{18} + C_2H_4$$

decane → octane + ethene

Figure 4.4 Cracking decane could produce octane (used in petrol) and ethene (used to make poly(ethene) plastic). However, if the decane molecule was broken in a different place, other different products could be made

Now test yourself

9 Explain why it is necessary to crack some hydrocarbon compounds.
10 Explain why cracking can be described as a 'thermal decomposition' reaction.
11 Suggest a use for the alkanes produced from cracking.
12 Suggest a use for the alkenes produced from cracking.

Answers on page 129

Alkanes

Alkanes as a homologous series

REVISED ☐

Alkanes are a **homologous series** of **hydrocarbon** compounds. This means that they are a 'family' of compounds which have similar **chemical properties** and a trend in **physical properties**. The simplest alkanes have an unbranched chain of carbon atoms, which is surrounded by hydrogen atoms. All the carbon atoms in an alkane are joined by single covalent bonds. This means that we describe the alkanes are being **saturated** hydrocarbons.

As the chain of carbon atoms lengthens from one alkane to the next alkane in the series, the molecular formula increases by one carbon atom and two hydrogen atoms. The first four alkanes are shown in Table 4.1. The names of all alkanes end in '-ane'. The fifth alkane is called pentane.

Revision activity

Devise a mnemonic to help you remember the names of the first five alkanes. This method might also help you to remember the names of alkenes, alcohols and carboxylic acids.

Alkanes can be described using a general formula, which is C_nH_{2n+2}. This means that if you know the number of carbon atoms in an alkane (n), you can work out the number of hydrogen atoms by doubling it and then adding 2.

Example

Deduce the molecular formula of heptane, which has seven carbon atoms.
- If $n = 7$, doubling it and adding 2 gives 16.
- So the formula of heptane must be C_7H_{16}.

Saturated: A hydrocarbon compound that contains only C–C single bonds. (Unsaturated hydrocarbons contain at least one C=C double bond.)

Alkane: A type of hydrocarbon in which the carbon atoms are all bonded by single C–C bonds.

Homologous series: A 'family' of compounds that has similar chemical properties, a trend in physical properties and whose members' chemical formulae differ by the addition of CH_2.

Hydrocarbon: A compound that is made from hydrogen and carbon atoms **only**.

Chemical properties: Descriptions of the ways that a substance characteristically behaves in chemical reactions. For example, how reactive it is with oxygen, water, or acids.

Physical properties: Descriptions of the ways that a substance characteristically behaves when it is not reacting with other chemicals. For example, its melting point, electrical conductivity, whether it is brittle, strong, hard, etc.

In hydrocarbon compounds, carbon always forms four covalent bonds and hydrogen always forms one covalent bond. You should use these rules to check all your drawings of hydrocarbon molecules.

Table 4.1 The first four alkanes

Name	Methane	Ethane	Propane	Butane
Molecular formula	CH_4	C_2H_6	C_3H_8	C_4H_{10}
Structural formula	CH_4	CH_3CH_3	$CH_3CH_2CH_3$	$CH_3CH_2CH_2CH_3$
Displayed formula				
Dot-and-cross diagram				
Molecular model (grey balls for carbon, blue for hydrogen)				

Properties of alkanes

REVISED

All alkanes are insoluble in water. The alkanes are chemically unreactive, but they do make very good fuels because they release a lot of energy when they are burned.

The longer the carbon chain in an alkane:

- The harder it is to set the alkane alight.
- The more energy is released when it burns.
- The stronger the intermolecular forces of attraction.
- The higher the boiling point.

Reaction with halogens

REVISED

Alkanes react with chlorine (Cl_2) and bromine (Br_2) when exposed to ultraviolet light. In this process, the UV light breaks apart the atoms in the halogen molecule. One of the halogen atoms then **substitutes** for (swaps places with) one of the hydrogen atoms on the alkane. An example of this is shown in Figure 4.5.

Substitution reaction: A reaction in which one atom swaps place with another atom in a molecule.

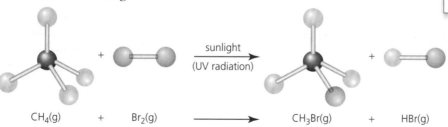

$CH_4(g)$ + $Br_2(g)$ ⟶ $CH_3Br(g)$ + $HBr(g)$

Figure 4.5 The reaction of methane with bromine, when exposed to ultraviolet light

Exam practice answers and quick quizzes at **www.hoddereducation.co.uk/myrevisionnotesdownloads**

Write a symbol equation to show the reaction of chlorine with ethane.

- Ethane is the second alkane, so it has two carbon atoms. Applying the general formula of C_nH_{2n+2}, ethane must have a molecular formula of C_2H_6.
- The chlorine molecule is diatomic, so it has the formula Cl_2.
- One of the hydrogen atoms on the ethane is substituted by a chlorine atom.
- $C_2H_6 + Cl_2 \rightarrow C_2H_5Cl + HCl$

Revision activity

Choose an alkane from Table 4.1 and choose either chlorine or bromine as the other reactant. Deduce the product that will be made when they react.

Exam tip

In these substitution reactions, HCl or HBr is always produced as a by-product.

Now test yourself

TESTED

13 State the name of the alkane which has five carbon atoms and deduce its molecular formula.
14 Which of the following alkanes will have the lowest boiling point: hexane, propane, ethane, butane? Explain your answer using ideas about intermolecular forces.
15 Draw the displayed formula of butane.
16 Write a symbol equation for the reaction of propane with bromine, when it is exposed to ultraviolet light.

Answers on page 129

Alkenes

Alkenes as a homologous series

REVISED

Alkenes are a homologous series of hydrocarbons which contain at least one C=C double bond. This means that the alkenes are described as **unsaturated** hydrocarbons.

The first three alkenes are shown in Table 4.2. You will see that their names are very similar to the corresponding alkanes, but that all alkenes have names that end in '-ene'. You will also see that there is no alkene containing only one carbon atom. This alkene would be impossible because there cannot be a double bond unless there are at least two carbon atoms.

Alkene: A type of hydrocarbon which contains at least one C=C double bond.

Unsaturated: A hydrocarbon compound that contains at least one C=C double bond.

Table 4.2 The first three alkenes

Name	Ethene	Propene	Butene
Molecular formula	C_2H_4	C_3H_6	C_4H_8
Structural formula	$CH_2{=}CH_2$	$CH_2{=}CHCH_3$	$CH_2{=}CHCH_2CH_3$
Displayed formula	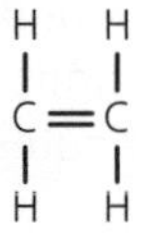	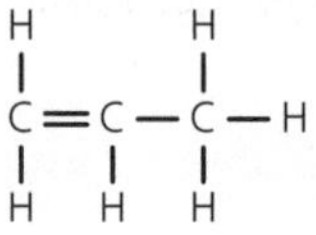	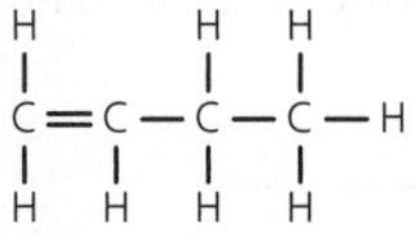

Like alkanes, alkenes have a general formula. This general formula is C_nH_{2n}.

Example

Deduce the molecular formula for pentene, which has five carbon atoms.

- If $n = 5$, doubling it gives 10.
- So the formula of pentene must be C_5H_{10}.

Exam tip

When drawing the displayed formulae of the alkenes, carefully check that each C atom has only made four bonds. It is very easy to make the mistake of filling in 'gaps' in their structures with H atoms, but that causes some C atoms to make five bonds, which is impossible.

Isomerism in the alkenes

REVISED

Isomers are molecules that share the same molecular formula but have a different arrangement of atoms. Because the molecules have a different structure, they usually have different physical and chemical properties.

Isomers: Molecules that have the same molecular formula, but a different arrangement of atoms.

In alkenes that have four or more carbon atoms, the double bond can be positioned in more than one place in the carbon chain. This means that butene has two isomers, which are shown in Figure 4.6.

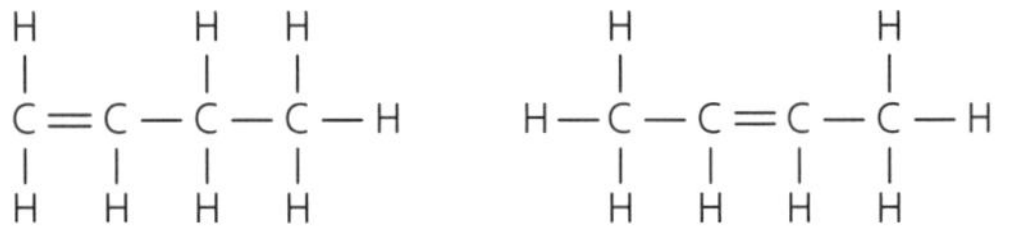

but-1-ene but-2-ene

Figure 4.6 The displayed formulae of the two isomers of butene

You can see that the two isomers of butene both have the same molecular formula, C_4H_8, yet they have different structures. The number which is inserted into the middle of the name tells you which carbon atom the double bond starts at. In but-1-ene, the C=C bond is at the end of the carbon chain, on the 1st C atom. In but-2-ene, the C=C bond starts on the 2nd C atom. The counting can begin at either end of the molecule, which means there is no such compound as but-3-ene (because that would be the same compound as but-1-ene).

Revision activity

Draw a blank table like Table 4.2 but include space for but-1-ene and but-2-ene. Now fill it in from memory. Check to see if you got everything right.

Reactions of alkenes

REVISED

The C=C part of an alkene is an example of a **functional group**. This means that it is a part of every alkene molecule, and it is this part of the molecule that determines the reactions of the homologous series. Each homologous series (e.g. alkenes, alcohols, carboxylic acids, esters) has its own functional group. You will learn more about them in the following sections.

The C=C functional group makes alkenes much more reactive than alkanes. One of the two bonds in the C=C double bond is likely to break, which allows one extra atom to be added on to each of the two carbons. These kinds of reactions are called **addition reactions** because new atoms are added.

One example of an addition reaction is when an alkene reacts with bromine, Br_2. The product of this reaction is a dibromoalkane. This name means that the molecule contains two bromine atoms and no longer has a

Functional group: A part of an organic molecule which is present in all members of a homologous series, and which determines how the compounds in that homologous series react with other substances. For example, the functional group of an alkene is C=C.

Addition reaction: A reaction in organic chemistry when two molecules combine to make one product. Addition reactions require one of the molecules to have a double bond.

C=C double bond (hence the ending -ane and not -ene). The addition of bromine to ethene is shown in Figure 4.7.

ethene   bromine   dibromoethane

Figure 4.7 The reaction of ethene with bromine

The reaction of bromine with an alkene can be used to test for the presence of a C=C double bond in an organic compound. In this test a solution of bromine in water is used. This solution is orange and it is usually called **bromine water**. If the hydrocarbon to be tested is a gas it is bubbled through the bromine water. If the hydrocarbon is a liquid, it is mixed with the bromine water and then shaken. If the hydrocarbon is unsaturated (i.e. if C=C bonds are present), the bromine water changes from orange to colourless, and we say that the bromine water has been decolourised.

Table 4.3 Testing for C=C double bonds using bromine water

	Saturated molecule	Unsaturated molecule
Add a few drops of orange bromine water, $Br_2(aq)$	No reaction – bromine water remains orange	Bromine water reacts and goes colourless

> **Typical mistake**
>
> You should say that the bromine water turns **colourless**, not **clear**. **Clear** means that you can see through it, but you can already see through the **clear orange** bromine water before you do the test.

Now test yourself

TESTED ☐

17 State the general formula for an alkene.
18 Draw the structure of propene.
19 Describe how to test for an alkene in the laboratory.
20 Draw the structure of the product created when bromine reacts with propene.

Answers on page 129

Alcohols

Alcohols as a homologous series

REVISED ☐

Alcohols are a homologous series of **organic compounds**, but they are not hydrocarbons because they all contain at least one oxygen atom. Alcohols all contain the functional group −OH. Alcohols have many uses: fuels, solvents and as raw materials in chemical reactions.

> **Organic compound:** A compound that contains C–H bonds. Alkanes, alkenes and alcohols are all examples of families of organic compounds.

The structures of the first four alcohols are shown in Table 4.4. You will notice that their names are similar to the name of the first four alkanes, but all end in '-ol'.

Table 4.4 The first four alcohols

Name	Methanol	Ethanol	Propanol (propan-1-ol)	Butanol (butan-1-ol)
Molecular formula	CH_3OH	C_2H_5OH	C_3H_7OH	C_4H_9OH
Structural formula	CH_3OH	CH_3CH_2OH	$CH_3CH_2CH_2OH$	$CH_3CH_2CH_2CH_2OH$
Displayed formula				

Manufacturing ethanol

REVISED

Ethanol can be produced by fermentation at low temperatures or by the direct hydration of ethene at high temperatures.

Fermentation

Ethanol can be made from glucose (a type of sugar) using a process called **fermentation**. In this process we use yeast. Yeast is a single-celled fungus which contains **enzymes** that convert the carbon, hydrogen and oxygen atoms in glucose ($C_6H_{12}O_6$) into ethanol and carbon dioxide. Fermentation has been used for thousands of years to produce alcoholic drinks. The word and symbol equations for fermentation are:

glucose → ethanol + carbon dioxide

$$C_6H_{12}O_6 \rightarrow 2C_2H_5OH + 2CO_2$$

The conditions required for fermentation are:
- an optimum temperature of about 30 °C
- an absence of air
- a mixture of sugar (e.g. glucose) and yeast in water.

If the temperature is too low, the reaction will either occur slowly or simply stop. If it is too high, the enzymes from the yeast become **denatured** (their shape changes permanently). This change in shape stops the enzymes from working and prevents fermentation from taking place.

If oxygen is present during fermentation, the ethanol will be **oxidised** to produce ethanoic acid, CH_3COOH. This is the acid that is present in vinegar, and it gives the alcoholic drink a sharp and unpleasant taste.

> **Denatured:** A permanent change in the shape of a protein molecule (e.g. an enzyme), usually caused by high temperature. A denatured enzyme will no longer work.
>
> **Oxidation:** A reaction in which an element or compound reacts with oxygen (or gains oxygen atoms).

> **Fermentation:** A biochemical process that releases energy from a sugar compound in the absence of oxygen.
>
> **Enzymes:** Proteins which are made by living organisms to speed up chemical reactions. Enzymes are therefore biological catalysts.

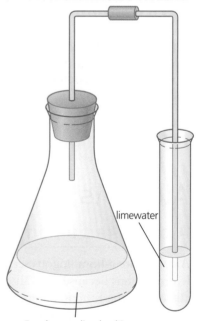

limewater

5 g glucose dissolved in 50 cm³ water + 1 spatula measure of yeast

Figure 4.8 Fermentation of glucose to make ethanol and carbon dioxide in the laboratory. The carbon dioxide turns the limewater cloudy

Hydration of ethene

Ethanol can also be made by reacting ethene with steam. This is an example of an addition reaction. As the water molecule (in steam) is what is added to the ethene, this is also often called a **hydration** reaction. The equations are:

ethene + water ⟶ ethanol

$$\underset{\underset{H}{\diagup}{\overset{\overset{H}{\diagdown}}{C}}=\underset{\underset{H}{\diagdown}}{\overset{\overset{H}{\diagup}}{C}} + H_2O \longrightarrow H-\overset{\overset{H}{|}}{\underset{\underset{H}{|}}{C}}-\overset{\overset{H}{|}}{\underset{\underset{H}{|}}{C}}-OH$$

The conditions needed for the hydration of ethene are:
● a temperature of about 300 °C
● a pressure of about 60–70 atmospheres (atm)
● a phosphoric acid catalyst.

Revision activity

Make a summary and comparison table for the two methods of producing ethene. Include the raw materials, conditions, equations and pros and cons.

Oxidation reactions of ethanol

REVISED

Ethanol can be oxidised during fermentation if air is present. This is called **microbial oxidation** because it takes place in the presence of microbes (yeast). Ethanoic acid is produced.

Ethanol can also be oxidised by heating it with potassium dichromate(VI) in dilute sulfuric acid. The product is ethanoic acid, CH_3COOH. The potassium dichromate(VI) is the oxidising agent in this reaction so it provides oxygen atoms to the ethanol molecules. During the reaction, the potassium dichromate(VI) solution turns from orange to green as the chromium atoms become reduced.

Alcohols are often used as fuels, and their combustion reactions are another example of oxidation. For example, the complete combustion of ethanol can be written as:

ethanol + oxygen → carbon dioxide + water

$$C_2H_5OH + 3O_2 \rightarrow 2CO_2 + 3H_2O$$

Typical mistake

When balancing combustion reactions for alcohols, it is very easy to forget to count the oxygen atom on the left-hand side in the formula for the alcohol. Make sure you double check when writing these reactions.

Now test yourself

TESTED

21 State the functional group present in all alcohols.
22 Draw the displayed formula for butanol.
23 State the conditions needed for fermentation.
24 Write a balanced symbol equation for the hydration of ethene. Explain why this reaction is an addition reaction.

Answers on page 129

Carboxylic acids

Carboxylic acids as a homologous series

REVISED

Carboxylic acids are a homologous series of organic compounds that all contain the functional group −COOH. The atoms in this functional group are bonded together as shown in Figure 4.9.

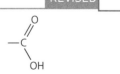

Figure 4.9 The functional group present in all carboxylic acids

The structures of the first four carboxylic acids are shown in Table 4.5. You will notice that their names are similar to the name of the first four alkanes, but all end in '-oic' acid. The most familiar carboxylic acid is probably ethanoic acid, and vinegar is an aqueous solution of this type of acid.

Table 4.5 The first four carboxylic acids

Name	Methanoic acid	Ethanoic acid	Propanoic acid	Butanoic acid
Molecular formula	HCOOH	CH_3COOH	C_2H_5COOH	C_3H_7COOH
Structural formula	HCOOH	CH_3COOH	CH_3CH_2COOH	$CH_3CH_2CH_2COOH$
Displayed formula				

Reactions of carboxylic acids

REVISED

Carboxylic acids are weak acids, which means that some of their molecules break apart to release hydrogen ions in a **reversible reaction** (see page 98). The relevant equations for ethanoic acid and propanoic acid are:

$$CH_3COOH \rightleftharpoons CH_3COO^- + H^+$$

$$CH_3CH_2COOH \rightleftharpoons CH_3CH_2COO^- + H^+$$

Carboxylic acids will react with some metals to produce a salt and hydrogen gas. The ending of the name of the salt produced is derived from the name of the carboxylic acid. These salts will contain the negative ion which is formed when the acid molecule loses its hydrogen ion. For example:

> **Reversible reaction:** A chemical change which can occur both ways; the products can turn back into the reactants, and vice versa.

sodium + ethanoic acid $\rightarrow$ sodium ethanoate + hydrogen

$$2Na + 2CH_3COOH \rightarrow 2CH_3COONa + H_2$$

Carboxylic acids will also react with metal carbonates to produce a salt, water and carbon dioxide. The name of the salt can be deduced from the metal in the carbonate and the name of the acid reacting. For example:

lithium carbonate + propanoic acid $\rightarrow$ lithium propanoate + water + carbon dioxide

$$Li_2CO_3 + 2CH_3CH_2COOH \rightarrow 2CH_3CH_2COOLi + H_2O + CO_2$$

Now test yourself

TESTED

25 Suggest the pH of a solution of propanoic acid.
26 Name the carboxylic acid with one carbon atom in each of its molecules.
27 Draw the structure of butanoic acid.
28 Write a word equation for the reaction of ethanoic acid with lithium.

Answers on page 129

Esters

Esters as a homologous series

REVISED

Esters are a homologous series of organic compounds that all contain the functional group shown in Figure 4.10.

Esters are volatile compounds that often have characteristic fruity smells so they are often used in perfumes and food flavourings. For example, one ester smells like apples and another smells like pineapples.

Figure 4.10 The functional group present in all esters

Making and naming esters

REVISED

Esters can be made by reacting a carboxylic acid with an alcohol. Water is produced in what is known as a **condensation reaction**. The name of the ester is derived from the alcohol (which gives the first part of the ester's name) and also from the name of the acid (which gives the second part). For example:

butanol + ethanoic acid $\rightleftharpoons$ butyl ethanoate + water

$$CH_3CH_2CH_2CH_2OH + CH_3COOH \rightleftharpoons CH_3CH_2CH_2CH_2OCOCH_3 + H_2O$$

These reactions are reversible, so a dynamic equilibrium will be established and the reaction mixture will contain reactants and products. The ester will need to be separated from the mixture. A concentrated acid catalyst is usually used to speed up the time taken for a dynamic equilibrium to be reached.

> **Condensation reaction:** A reaction in which two molecules combine and a small molecule is made as a by-product. The small molecule is usually water.

> **Revision activity**
>
> Choose an alcohol from Table 4.4 and a carboxylic acid from Table 4.5. Deduce the name of the ester that they will make when they react, then check your answer. You can repeat this activity several times with different options selected.

Required practical

Prepare a sample of an ester such as ethyl ethanoate

Method

1 $1\,cm^3$ of ethanol (caution: flammable) was placed into a test tube.
2 $1\,cm^3$ of concentrated ethanoic acid (caution: corrosive) was added to the test tube.
3 Three drops of concentrated sulfuric acid (caution: corrosive) were added to the test tube to act as a catalyst.
4 The neck of the test tube was sealed with cling film.
5 The test tube was placed in a warm water bath for 10 minutes.
6 The contents of the test tube were poured into a beaker containing $50\,cm^3$ of sodium carbonate solution. This neutralised any remaining acid.
7 The oily droplets on the surface of the solution contained the ester.

Drawing esterification reactions

REVISED

The equations below show the condensation reaction between ethanoic acid and ethanol. The dotted line shows the atoms which become the water molecule during this reaction.

ethanoic acid + ethanol $\rightleftharpoons$ ethyl ethanoate + water

Similar diagrams can be drawn for the formation of different esters. For example:

propanoic acid + butanol ⇌ butyl propanoate + water

$CH_3CH_2COOH + HOCH_2CH_2CH_2CH_3 \rightleftharpoons CH_3CH_2COOCH_2CH_2CH_2CH_3 + H_2O$

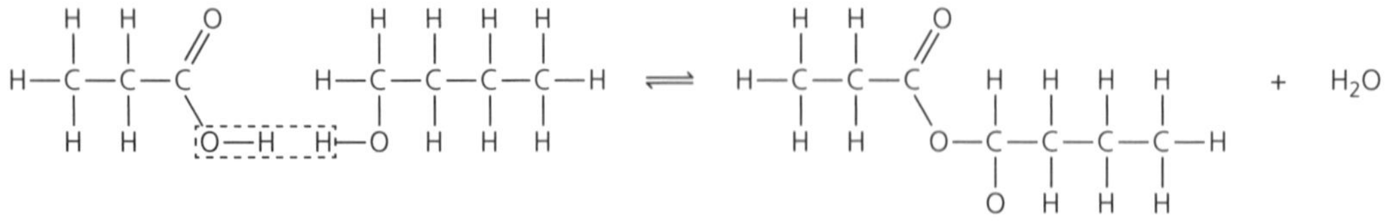

Now test yourself

TESTED

29 Draw the functional group which is present in all esters.
30 Write a word equation for the reaction of methanol with ethanoic acid.
31 Deduce the structural formula of the ester that will be produced from ethanol and propanoic acid.
32 Draw the displayed formula of the ester that will be produced from propanol and ethanoic acid.

Answers on page 129

Revision activity

Make flashcards for each of the homologous series you have learned about. Include the functional group, an example of a specific compound, and any other relevant details.

Addition polymers

Making a polymer

REVISED

Polymers are very long molecules made up from smaller molecules that are linked up in a chain. The smaller molecules are called **monomers**. The chemical reaction in which thousands of monomers are joined together to produce a polymer is called polymerisation. Polymerisation is the process that produces plastics. During **addition polymerisation**, no by-products are made. Addition polymerisation requires each monomer to have a C=C double bond.

The name of the polymer produced is derived from the name of the monomer, and by adding the prefix 'poly-'. For example, the polymer made by using ethene as the monomer is called poly(ethene). Similarly, polymerising propene monomers produces poly(propene).

Polymer: A very long molecule which is made of repeating smaller sections.

Monomer: A small molecule which can be added together with other monomers to produce a polymer.

Addition polymerisation: The process of combining unsaturated polymers (which contain a C=C double bond) to produce a polymer without any by-products.

Example

State the name of the polymer made from chloroethene.

- Poly(chloroethene)

The addition polymerisation of ethene monomers can be represented in the following way:

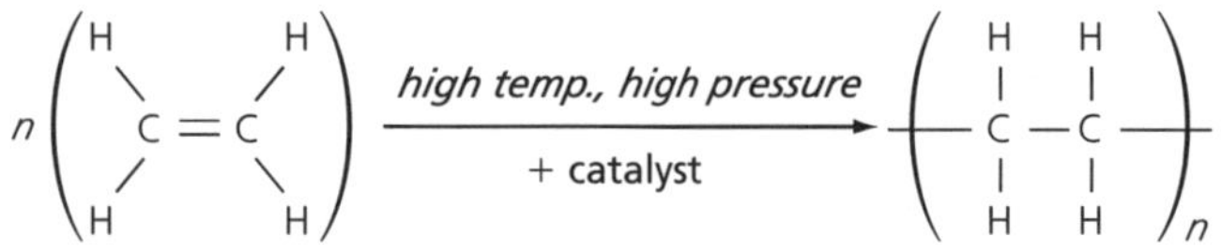

In this diagram, '*n*' represents a very large number. Typically, several thousand monomers react to produce a polymer which is thousands of carbon atoms long.

Typical mistake

Because the name of an addition polymer comes from the name of the alkene monomer used, it sounds like the polymer contains C=C double bonds, but this is not the case. You need to remember that the double bond from the monomer turns into a C–C single bond when the polymer is made.

Different polymers from different monomers

Starting with a different monomer will produce a polymer with a different structure and therefore it will also have different properties. These properties will make it more suitable for a particular application than another polymer might be. Table 4.6 shows the polymer structures made from four different monomers.

Exam tip

You need to be able to deduce the structure of a polymer if you are given the structure of a monomer, and vice versa, as it is a common exam question.

Table 4.6 The polymers made from different monomers

Monomer name and structure	Polymer name and structure	Uses
Ethene	Poly(ethene)	Plastic bottles, bags, food containers
Propene	Poly(propene)	Plastic crates and ropes
Chloroethene	Poly(chloroethene) (PVC)	Guttering, window frames, insulation for electrical wires
Tetrafluoroethene	Poly(tetrafluorethene) (PTFE/Teflon)	Coating for non-stick saucepans

Revision activity

Draw the structure and write the name of each of the monomers from Table 4.6 on separate cards. On the reverse, draw the structure and name of the relevant polymer. Shuffle the cards, then choose one. From memory, draw what is on the reverse of the card, then flip it over to check if you were right.

Problems with addition polymers

Addition polymers have many desirable properties, which make them useful in a wide range of applications in everyday life. For example, they are usually strong, flexible, waterproof and unreactive. However, they also have problems. For example, if addition polymers are burned, they often produce toxic gases which can be harmful for living things, including humans.

Another problem is that their low chemical reactivity makes it very hard for microorganisms (e.g. bacteria and fungi) to break them down when they are disposed of. Addition polymers are therefore not usually **biodegradable**, so they may last for hundreds of years in landfill sites.

Biodegradable: A substance or object that can be broken down through the action of living organisms in the environment.

Now test yourself

TESTED

33 State the functional group which is required in all addition monomers.
34 What is the name of the polymer which is produced from butene?
35 What is the name of the monomer which is used to produce poly(phenylethene)?
36 Draw the structure of poly(ethene).

Answers on page 129

Condensation polymers

Making polyesters

We have already seen how esters are made as part of a **condensation reaction** between an alcohol and a carboxylic acid. During a condensation reaction, a small molecule is produced as a by-product. In the case of an ester, the by-product is water (see page 117).

Polyesters are a type of **synthetic polymer** which are normally made up of two different monomers. The first type of monomer has a carboxylic acid group at each end of the molecule, so it is called a **dicarboxylic acid**. The second type of monomer has an alcohol group at each end of the molecule, so it is called a **diol**.

> **Dicarboxylic acid:** An organic molecule that has two carboxylic acid (–COOH) groups.
>
> **Diol:** An organic molecule that has two alcohol (–OH) groups.

> **Condensation reaction:** A reaction in which two molecules combine and a small molecule is produced as a by-product. This small molecule is usually water.
>
> **Synthetic polymer:** A polymer which is made by chemists for its useful properties. Poly(ethene) and nylon are examples of synthetic polymers.

The reaction of ethanedioic acid with ethanediol to make a polyester is shown in Figure 4.11:

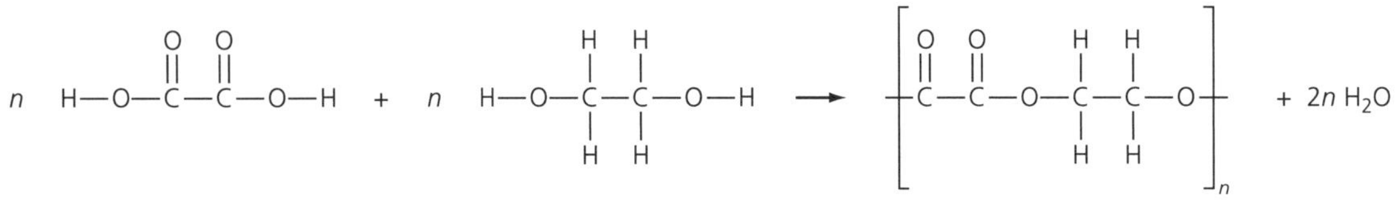

Figure 4.11 The formation of a polyester from ethanedioic acid and ethanediol

Example

Write the structural formula of the repeat unit of the polyester formed from pentanedioic acid ($HOOCCH_2CH_2CH_2COOH$) and butanediol ($HOCH_2CH_2CH_2CH_2OH$).
- Remove the HO from both ends of the dicarboxylic acid monomer: $-OCCH_2CH_2CH_2CO-$
- Remove the H from both ends of the diol: $-OCH_2CH_2CH_2CH_2O-$
- Start with the dicarboxylic acid and then add the diol onto the end:

$[-OCCH_2CH_2CH_2COOCH_2CH_2CH_2CH_2O-]_n$

Example

Draw the displayed formula of the repeat unit of the polyester formed from butanedioic acid and propanediol. The displayed structures of the monomers are shown below.

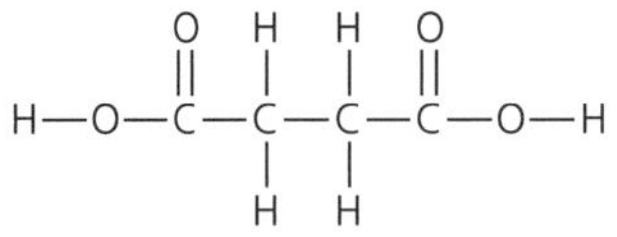

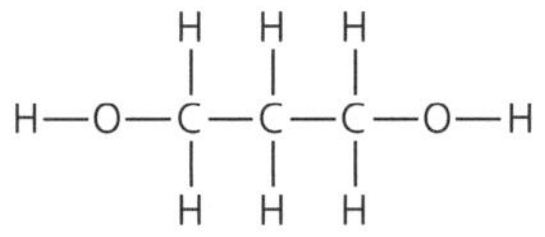

- Remove the HO from both ends of the dicarboxylic acid monomer:

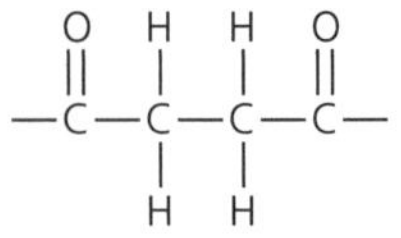

- Remove the H from both ends of the diol:

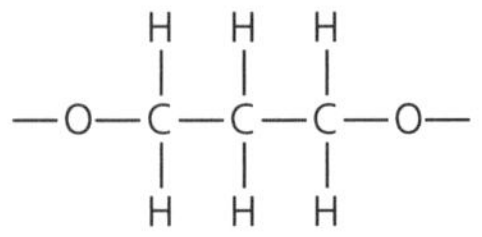

- Start with the dicarboxylic acid and then add the diol onto the end:

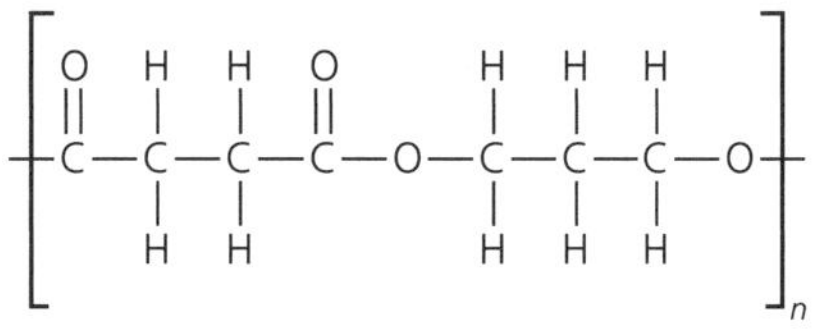

Polyesters have a range of useful properties which mean they can be used in a variety of applications. For example, they are used in fabrics for clothing, to make water bottles, and in varnishes.

Biopolyesters

REVISED

As discussed on page 119, polymers are usually unreactive and this makes it very hard for microorganisms to break down plastics when they are disposed of. However, scientists are now researching ways of making plastics that are biodegradable. One example of this research is seen in **biopolyesters**. Scientists are interested in biopolyesters because many naturally occurring polymers (those that are made by living organisms) are actually examples of polyesters. These include starch and cellulose, both of which can be broken down by living organisms.

> **Exam tip**
>
> You need to be able to draw out the structural and displayed formulae of a polyester if you are given the formulae of the two monomers in the exam.

> **Biopolyester:** A polymer that contains ester linkages between monomers, which can be broken down by living organisms, and so is biodegradable.

Now test yourself

TESTED

37 Describe the difference between an addition polymer and a condensation polymer.
38 Name two types of monomers that can be reacted together to produce a polyester.

Answers on page 129

> **Revision activity**
>
> Draw up a comparison table for addition polymerisation and condensation polymerisation. Include the types of monomer used, a few examples of polymer names and their repeat units, structures, and their uses.

Summary

- Fractional distillation of crude oil separates it into fractions which are mixtures of hydrocarbons with similar boiling points.
- Longer hydrocarbons have stronger intermolecular forces of attraction and so they have a higher boiling point.
- Complete combustion of a hydrocarbon releases the maximum amount of energy and produces water plus carbon dioxide.
- Incomplete combustion occurs when there is insufficient oxygen, and it releases less energy. Water is produced, along with poisonous carbon monoxide or particulate carbon (which can cause health problems).
- Sulfur impurities in fossil fuels are oxidised to SO_2 when the fuel is burned. The SO_2 dissolves in rain water to cause acid rain.
- High temperatures in car engines result in the production of nitrogen oxides which also cause acid rain.
- Cracking long hydrocarbons into shorter, more useful hydrocarbons is done to more accurately match supply with demand. A shorter alkane and a shorter alkene are produced.
- Alkanes are saturated hydrocarbons in which all C atoms are joined by single bonds (C–C).

- Alkenes are reactive unsaturated hydrocarbons which have at least one C=C double bond.
- Alcohols all contain the functional group –OH.
- Ethanol can be made through fermentation using yeast or by direct hydration of ethene.
- Carboxylic acids all contain the functional group –COOH. These are weak acids which react with some metals and all metal carbonates.
- Esters are made from an alcohol and a carboxylic acid and all contain the functional group –COO–. Esters have characteristic smells which mean they are used for perfumes and food flavourings.
- Addition polymers can be made from alkene monomers. Addition polymers include polyethene which is used for plastic bags and buckets. Addition polymers are not usually biodegradable, so they can cause environmental problems.
- Condensation polymers include polyesters. When they are made, a small molecule is produced every time two monomers join. Polyesters are usually made from a dicarboxylic acid and a diol.
- Biopolyesters can be broken down by living organisms so they are biodegradable.

Exam practice

1 Crude oil is a non-renewable resource that is a mixture of hydrocarbon compounds. It can be separated into useful substances using fractional distillation.

   a State what is meant by the term 'hydrocarbon'. [2]

   b The figure below shows the equipment used in the industrial fractional distillation of crude oil.

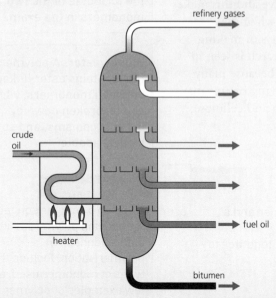

     i On a copy of the diagram, draw an X inside the fractionating tower where the temperature is the hottest. [1]

     ii Label the missing fractions that are separated from crude oil in this process and state a use for each one. [6]

   c Describe how fractional distillation is used to separate crude oil into useful substances. You should use ideas about relevant physical properties and forces between molecules in your answer. [4]

   d More fuel oil is obtained from crude oil than is needed, so there is a surplus of unwanted fuel oil which cannot be sold.

     i State the industrial process that is used to convert unwanted fuel oil into useful hydrocarbons. [1]

     ii Explain how this process helps to match supply with demand. [3]

2 A fuel is a substance that is burned to release energy in a combustion reaction. The products of combustion depend on the fuel that is being burned and the conditions during combustion.

   a Write a balanced symbol equation for the complete combustion of the hydrocarbon fuel methane. [2]

   b Write a word equation for the incomplete combustion of methane. [1]

   c State what condition is required during the combustion of methane for incomplete combustion to occur. [1]

   d Identify **three** disadvantages of the incomplete combustion of methane, compared with the complete combustion of octane. [3]

   e Chemists ensure that all sulfur impurities are removed from fuels before they are used in cars. Explain why this is important for preventing pollution. [3]

   f Car exhaust gases contain nitrogen oxides which are another pollutant.

     i State an environmental problem caused by nitrogen oxides. [1]

     ii Explain how nitrogen oxides are produced in a car engine. [2]

3 Alcohols are an important homologous series of organic compounds because they are useful fuels and solvents and can be used in chemical reactions to produce other useful compounds.

   a State the molecular formula of ethanol. [1]

   b Draw the structure of propanol (propan-1-ol). [1]

   c Ethanol can be produced by fermentation or by the hydration of ethene. Identify and explain which method would be best for the small-scale production of ethanol in a remote location in a tropical country which has limited supplies of crude oil. [5]

Answers and quick quizzes online

ONLINE

Now test yourself answers

1 Principles of chemistry 1

1 Atoms are the smallest particles of an element and they are neutral; ions are small particles that have either a positive or negative charge; molecules are clusters of non-metal atoms that are chemically bonded together and they are neutral.

2 Solvent

3 g of solute per 100 g of solvent

4 No more solute can dissolve in a saturated solution.

5 Condensation

6

	Solid	Liquid	Gas
Arrangement of particles	Regularly arranged, touching neighbours	Randomly arranged, mostly touching neighbours	Widely spaced
Movement of particles	Vibrating	Can move over each other	Moving fast

7 Evaporation

8 Sublimation

9 K

10 Compound

11 Two carbon atoms, four hydrogen atoms, two oxygen atoms

12 Because air is a mixture

13 Filtration

14 Distillate

15 As the mixture contains more than two liquids, you must use fractional distillation instead.

16 The cold water should go into the lower connector on the condenser and leave at the upper connector.

17 Crystallisation

18 Residue

19 The solvent/water/chemical which moves through the stationary phase and carries the individual substances from the mixture

20 The red pigment is insoluble in water / not attracted at all to the water.

21 The R_f value will always be less than 1 because the distance moved by the solvent will always be more than the distance moved by the chemical in the formula:

$$R_f \text{ value} = \frac{\text{distance moved by chemical}}{\text{distance moved by solvent}}$$

22 Protons and neutrons

23 1

24 Zero/no charge/neutral

25 The number of protons in an atom

26 You subtract the atomic number from the mass number.

27 20

28 3 protons, 4 neutrons, 3 electrons

29 19 protons, 20 neutrons, 19 electrons

30 $^{12}_{6}C$ has 6 protons, 6 neutrons and 6 electrons.

$^{14}_{6}C$ has 6 protons, 8 neutrons and 6 electrons.

They are isotopes because they have the same number of protons and different numbers of neutrons.

31 They have the same chemical properties and they react the same way.

32 Relative atomic mass $= \dfrac{(10 \times 20) + (11 \times 80)}{100}$

$= \dfrac{200 + 880}{100} = 10.8$

33 The first shell can hold two electrons, and the second shell can hold eight electrons.

34 2,8,5

35 Aluminium

36 2,8

37 Sodium atom: 11 protons, 12 neutrons, 11 electrons; its electron configuration is 2,8,1.

A sodium ion must have lost the single electron in its outer shell because it is in group 1. So it has 11 protons, 12 neutrons and 10 electrons; the electron configuration of Na^+ is 2,8.

38 The nucleus of a plant cell is much larger as it contains millions of atoms, each with its own nucleus.

39 Carbon-12

40 A_r

41 Periodicity

42 John Newlands

43 Group 0/the noble gases

44 Period

45 Halogens

46 Group 1

47 Potassium

48 Because they have a full outer shell of electrons and do not need to gain, lose or share electrons to become stable.

49 Radon (although xenon is also acceptable as radon is not discussed in the section)

50 The boiling point increases.

51 Helium

1 Principles of chemistry 2

1 27

2 $2 \times (6 \times 10^{23}) = 1.2 \times 10^{24}$

3 $(2 \times 27) + (3 \times 16) = 102$

4 $102\,g$

5 $204 \div 102 = 2\,mol$

6 Mass of oxygen atoms that reacted
   $= 39.75 - 31.75 = 8\,g$

	Copper	Oxygen
Reacting masses (g)	31.75	8
A_r	63.5	16
Moles of atoms = mass ÷ A_r	$31.75 \div 63.5 = 0.5$	$8 \div 16 = 0.5$
Simplified ratio	1	1
Empirical formula	CuO	

7 H_2S

8 CH_4

9 Calcium is in group 2, so it must have ions with a 2+ charge.

10 $CuCO_3$

11 $Zn(NO_3)_2$

12 Calcium carbonate
   → calcium oxide + carbon dioxide

13 Nitric acid + sodium hydroxide
   → sodium nitrate + water

14 $2NO + 2CO \rightarrow N_2 + 2CO_2$

15 $2Al(s) + Fe_2O_3(s) \rightarrow 2Fe(l) + Al_2O_3(s)$

16 $(2 \times 56) + (3 \times 16) = 160$

17 Moles of $Fe_2O_3 = \frac{480}{160} = 3\,mol$

   Moles of Fe = $2 \times$ mol of $Fe_2O_3 = 6\,mol\,of\,Fe$

   Mass of Fe = mol $\times A_r = 6 \times 56 = 336\,g$

18 Moles of $Fe_2O_3 = \frac{480}{160} = 3\,mol$

   Moles of $CO_2 = 3 \times$ mol of $Fe_2O_3 = 9\,mol\,of\,CO_2$

   Volume of CO_2 = mol of $CO_2 \times 24$
   $= 9 \times 24 = 216\,dm^3$

19 Percentage yield $= \frac{125}{150} \times 100 = 83\%$

20 Lithium iodide is ionic; sulfur dioxide is covalent; bromine is covalent; iron oxide is ionic.

21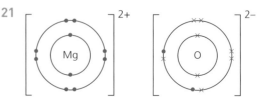

22 Na_2S

23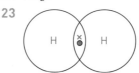

24 Conductors: iron, calcium, cobalt, graphite
   Insulators: iodine, diamond

25 No, because it is a molecular covalent substance.

26 No, because it is a molecular covalent substance.

27 Yes, because it is a molten ionic compound.

28 No, because solutions of molecular covalent substances do not conduct electricity.

29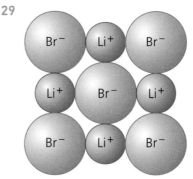

30 MgO will have a higher melting point because the Mg^{2+} ion has a higher charge than the Na^+ ion, and the O^{2-} ion has a higher charge than the F^- ion, so MgO will have stronger electrostatic forces of attraction.

31 NaF will have a higher melting point than KCl because the Na^+ ion is smaller than the K^+ ion, and the F^- ion is smaller than the Cl^- ion, so NaF will have stronger electrostatic forces of attraction.

32 In solid zinc chloride, the ions are fixed in place so they cannot move to carry a current. In molten zinc chloride, the ions are free to move and carry a current.

33 Pure water is made of molecules so it does not conduct, but when NaCl dissolves, the Na^+ and Cl^- ions can move to carry a current.

34 Substance X has a giant covalent structure.

35 Ionic; giant ionic lattice

36 Z is a non-metal. The low boiling point tells us that the oxide must have a simple molecular covalent structure, so Z must be a non-metal.

37 Molecule

38 Simple molecular covalent: bromine, hydrogen sulfide, carbon monoxide
Giant covalent: silicon, graphite

39 Diamond, because it has a giant covalent structure, whereas the other three have a simple molecular structure.

40 Butane is made from smaller molecules (molecules with a lower M_r) and therefore it has weaker intermolecular forces of attraction than hexane.

41 It has a high melting point because it requires a lot of energy to break the many strong covalent bonds between the carbon atoms.

42 Diamond is very hard, so it will cut through other substances, and it has a very high melting point, so it can withstand the heat caused by friction in the cutting process, without melting.

43 In graphite, each carbon atom is bonded to three others, leaving one electron spare. This electron is delocalised, so it can move through the structure to conduct electricity. In diamond, all four outer electrons on each atom are involved in bonding to four neighbouring atoms, so there are no delocalised electrons.

44 Similarities: carbon atoms are bonded to three other atoms; and the structures of both include hexagons.
Differences: graphite is made from layers, whereas C_{60} is made from spheres.

45 Similarities: carbon atoms are bonded to three other atoms; and they have a hexagonal arrangement of atoms.
Differences: graphene is made of a single sheet of atoms; nanotubes are made of a hollow cylinder/rolled sheet.

46 Any three of: high melting point, good conductor of electricity, good conductor of heat, high density, malleable

47 You could connect up a simple circuit containing a DC power supply (cell or powerpack) with either a bulb or ammeter in series with a sample of the solid substance. If a current flows (or the lamp lights), then the sample is either a metal or graphite. If it is a brittle black solid, it is probably graphite. Otherwise, it is a metal.

48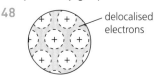
delocalised electrons

49 They are good conductors because delocalised electrons are able to move through the metallic lattice.

50 No, it is not a metal and it is ionic. Solid metals would conduct electricity but ionic substances do not conduct when solid because the ions are fixed in place. Once ionic substances are melted the ions are free to move and carry a current.

51 Simple molecular covalent substances do not contain ions so they cannot conduct electricity.

52 Anode

53 The lead ion is attracted to the cathode, where it gains electrons (it is reduced) to form a lead atom.

54 Hydrogen at the cathode; bromine at the anode

2 Inorganic chemistry

1 Rubidium hydroxide

2 They react similarly because they all have one electron in their outer shell which they lose when they react.

3 Potassium is more reactive because its outer electron is further from the nucleus, so it is held less strongly and lost more easily during the reaction with oxygen.

4 Lithium has a higher melting point than sodium or potassium.

5 A substance that is made up of molecules containing two atoms

6 It is likely to be a black solid.

7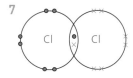
chlorine, Cl_2
Cl–Cl

8 Chlorine is a smaller molecule (has a lower M_r) than bromine, so it has weaker intermolecular forces of attraction, and therefore it requires less energy to overcome these forces.

9 Yes it will, because fluorine is more reactive than chlorine.

10 Bromine + potassium iodide
→ iodine + potassium bromide

11 $Cl_2 + 2LiBr \rightarrow Br_2 + 2LiCl$

12 $Cl_2 + 2I^- \rightarrow I_2 + 2Cl^-$

13 The bromine atoms are too large to be able to remove an electron from the chloride ions. The electrons are more strongly attracted to the chlorine nucleus than they would be to the bromine.

14 The copper (in copper oxide) has been reduced.

15 Carbon (C) is the reducing agent because it has removed the oxygen from the CuO.

16 Reduction (the Fe^{3+} is gaining electrons)

17 $2Br^- \rightarrow Br_2 + 2e^-$

18 Nitrogen, oxygen, argon, carbon dioxide

19 21%

20 The oxygen atoms react with the magnesium atoms to form solid magnesium oxide, so the volume of air decreases. Since the MgO does not take up any more volume than the Mg did, the overall volume decreases.

21 You can wedge some iron wool into the end of a measuring syringe, then invert the gas syringe and dip it into a water bath. You should then clamp it in this position. The volume of air in the measuring cylinder should be recorded.

   This set-up should be left for several days until the iron has fully rusted. After this period, the new volume of air in the measuring cylinder should be measured. You can then calculate the volume decrease and express this as a percentage of the original volume of air.

22 Insert a glowing splint into the location where you believe oxygen to be present. If the splint relights, the gas is oxygen.

23 Sodium + oxygen → sodium oxide

24 $2Mg + O_2 \rightarrow 2MgO$

25 Element X is a metal because the oxide of X has a high melting point, so it is likely to be ionic. When the oxide dissolves in water, it is also shown to be an alkali. This confirms that X was a metal.

26 Sulfuric acid + sodium carbonate → sodium sulfate + water + carbon dioxide

27 $CaCO_3 \rightarrow CaO + CO_2$

28 $2HCl + CuCO_3 \rightarrow CuCl_2 + H_2O + CO_2$

29 A gas that absorbs infrared radiation in the atmosphere, contributing to the greenhouse effect

30 Calcium + hydrochloric acid → calcium chloride + hydrogen

31 $Zn + H_2SO_4 \rightarrow ZnSO_4 + H_2$

32 $2Mg + O_2 \rightarrow 2MgO$

33 Tin is less reactive than magnesium because it cannot displace it.

34 Metal Y > metal Z > metal X

35 Oxygen and water present

36 A sample of distilled (fresh) water is placed in a test tube. A sample of sodium chloride solution (or actual sea water) is placed in another test tube. An iron or steel nail is placed into each test tube. Observations are made each day to see which one rusts the fastest, or to see which one rusts the most in a specific time.

37 Painting / galvanising

38 Sacrificial protection / galvanising / painting

39 Oil or grease

40 Any two from: oxygen / sulfur / carbon

41 A rock which contains enough metal or metal compound to make it economical to quarry or mine it

42 It is unnecessary and too expensive, when displacement with carbon is cheaper and just as effective.

43 Calcium is more reactive than carbon, so it cannot be displaced using carbon.

44 It can be extracted through displacement using carbon.

45 Carbon / carbon monoxide

46 It is added to remove impurities from the iron ore.

47 Carbon / carbon monoxide

48 Aluminium is more reactive than carbon, so it cannot be displaced by carbon.

49 Electrolysis

50 It is cheaper (requires less energy) to dissolve aluminium oxide into molten cryolite at approximately 1000 °C than it is to melt the aluminium oxide at approximately 2000 °C.

51 $Al^{3+} + 3e^- \rightarrow Al$

52 It is strong and hard so does not bend when used to cut food.

53 Aluminium and copper are both good conductors of heat and both have high melting points, and neither reacts with the chemicals in foods. However aluminium has a lower density, so it is easier to lift the pan when cooking.

54 An alloy is a mixture of a metal element with at least one other element (which is usually another metal).

55 Copper is a better conductor of electricity, so it is used in the wiring in the home. It has a higher density than aluminium, and the extra weight of the wires is not an issue in the home. However, thick, long cables made from copper would put too much strain on pylons, so the pylons would need to be very strong and that would make them very expensive.

56 3: colourless

6: colourless

7: colourless

9: pink

12: red

57 Magnesium + sulfuric acid
→ magnesium sulfate + hydrogen

58 Zinc oxide + hydrochloric acid
→ zinc chloride + water

59 $LiOH + HNO_3 \rightarrow LiNO_3 + H_2O$

60 Volumetric pipette

61 Pipette filler

62 So that the colour change of the indicator can be more easily seen

63 You stop adding the chemical when the indicator first changes colour permanently.

64 A hydrogen ion, H^+ (also called a proton)

65 An alkali is a solution made from a soluble base. A base might be insoluble, in which case it is not an alkali. All alkalis are bases, but not all bases are alkalis.

66 Hydroxide ion, OH^-

67 A proton donor

68 Soluble

69 Insoluble

70 Soluble

71 You would use cobalt chloride and sodium carbonate. To prepare cobalt carbonate you need to first mix appropriate volumes of the reactants together, e.g. $20 \, cm^3$ of each, if they have the same concentration (e.g. $0.1 \, mol/dm^3$). The next steps are to filter the products and rinse the residue. This should then be pressed between paper towels and dried in an oven.

72 You would use hydrochloric acid as the source of the chloride ion. To prepare calcium chloride, you would add calcium carbonate powder to hydrochloric acid a little at a time, with stirring, until the calcium carbonate is in excess. Next you would filter the excess calcium carbonate and discard. Finally you would evaporate half of the volume of the filtrate by heating, before allowing it to cool and crystallise. Crystals should then be removed and dried.

73 You would add small volumes of nitric acid to a solution of potassium hydroxide, with stirring. After each addition, you should remove a small sample of the mixture and test its pH. When the solution is neutral, evaporate half the volume. You should then let this cool and crystallise before removing the crystals and drying. (Alternative method: Use titration to find the volume of nitric acid required to exactly neutralise a measured volume of potassium hydroxide. Repeat the titration without the indicator and then evaporate, crystallise, and dry the crystals.)

74 The flame test indicates K^+.

The barium chloride test indicates SO_4^{2-}.

The compound is therefore potassium sulfate, K_2SO_4.

75 The sodium hydroxide test indicates Cu^{2+}.

The silver nitrate test indicates Br^-.

The compound is therefore copper bromide, $CuBr_2$.

76 You would split the sample of gas into two test tubes. The first sample should be tested with a burning splint. If it is hydrogen, it will give a squeaky pop and a small explosion. The second sample should be tested by inserting a glowing splint. If it is oxygen, the splint will relight.

77 You would hold a piece of damp universal indicator paper (or **red** litmus paper) near the top of the test tube. If the paper goes white the gas is chlorine. If the paper turns blue, the gas is ammonia.

3 Physical chemistry

1 Endothermic

2 Exothermic

3 Negative

4 To prevent heat loss from affecting the results

5 $Q = 150 \times 4.2 \times 23$

$= 14\,490\,J$

$= 14.49\,kJ$

Enthalpy change $= -14.49 \div 0.02$

$= -724.5\,kJ/mol$

6 Temperature change $= 11 - 23$

$= -12\,°C$

$Q = 30 \times 4.2 \times -12$

$= -1512\,J$

$= -1.512\,kJ$

Moles of $NH_4NO_3 = 2 \div 80$

$= 0.025$

Enthalpy change $= -(-1.512) \div 0.025$

$= 60.48\,kJ/mol$

(N.B. The value is positive because it is endothermic: the temperature decreased.)

7 Bond breaking occurs first. This process is endothermic.

8 kJ/mol

9

Bonds broken in reactants		Bonds made in products	
One H–H bond	1 × 436 = 436	Two H–Cl bonds	2 × 431 = 862
One Cl–Cl bond	1 × 242 = 242		
Total	678	Total	862

Enthalpy change = (energy in to break bonds) – (energy out when new bonds are made)
= 678 – 862
= –184 kJ/mol

10

Bonds broken in reactants		Bonds made in products	
Two H–H bonds	2 × 436 = 872	Four H–O bonds	4 × 464 = 1856
One O=O bond	1 × 498 = 498		
Total	1370	Total	1856

Enthalpy change = (energy in to break bonds) – (energy out when new bonds are made)
= 1370 – 1856
= –486 kJ/mol

11 You could measure mass change over time using a top pan balance. Alternatively you could measure the volume of gas over time using a gas syringe.

12 As it builds up the precipitate of solid sulfur will make the solution opaque, so you could measure the time taken for a cross to disappear when looking through the solution.

13 Average rate = 40 ÷ 20
= 2 cm³/s

14 The tangent extends to (0, 30) and (60, 70). The change in volume is 70 – 30 = 40 cm³; while the change in time is 60 – 0 = 60 s. Therefore the rate is calculated by 40 ÷ 60 = 0.67 cm³/s. (*N.B. Your answer might be a little different if your tangent was at a slightly different angle.*)

15 The minimum energy required by particles for a collision to result in a reaction between particles

16 You should have suggested a time shorter than 43 seconds.

17 Increasing the concentration means that there are more acid particles per unit volume. Therefore, successful collisions between particles are more frequent and the reaction will be quicker.

18 At lower temperatures (in the fridge), particles that could meet and cause decay have less energy so successful collisions are less frequent. As well as this, a lower proportion of collisions are successful because fewer particles have the required activation energy.

19 Iron is used as a catalyst to speed up the reaction by providing an alternative reaction pathway which has a lower activation energy.

20 Powder has a larger surface area.

21 Powders or gauzes have a large surface area which means collisions are more frequent.

22 The tangent for the large marble chips should go from approximately (0, 125.6) to (60, 118.0) which means a gradient (and therefore rate) of 7.6 ÷ 60 = 0.13. The tangent for the small marble chips should go from approximately (0, 125.6) to (20, 118.0) which means a gradient (and therefore rate) of 7.6 ÷ 20 = 0.38.

23 A reaction that can go both ways or a reaction where products can turn back into reactants

24 ⇌

25 Carbon monoxide + hydrogen ⇌ methanol

26 $CO(g) + 2H_2(g) \rightleftharpoons CH_3OH(g)$

27 Dynamic equilibrium is where a reversible reaction in a sealed container has reached the point when the forward and reverse reactions are occurring at the same rate, so the concentrations of the products and the reactants remain constant over time.

28 Increasing the pressure has no effect, because there are 2 moles of gas on both sides of the equation.

29 A low temperature should be used to produce the maximum amount of sulfur trioxide.

4 Organic chemistry

1 CH_4 and C_6H_6

2 Fractional distillation

3 Kerosene

4 As the length of the molecules increases, so does the viscosity.

5 Octane + oxygen → carbon dioxide + water

6 $C_3H_8 + 5O_2 \rightarrow 3CO_2 + 4H_2O$

7 Less energy is released; poisonous CO is produced; dirty soot/particulate carbon is produced, which causes breathing difficulties.

8 Removing sulfur prevents the formation of sulfur dioxide when the fuel is burned, and this reduces acid rain pollution.

9 Cracking is required to match the supply and demand for the hydrocarbons obtained from fractional distillation. For the shorter fractions, demand exceeds supply, but for the longer fractions, supply exceeds demand. Cracking converts longer hydrocarbons into smaller ones.

10 It is thermal decomposition as long hydrocarbon molecules are broken down using heat.

11 The alkanes can be used for petrol.

12 The alkenes can be used to make plastics.

13 Pentane, C_5H_{12}

14 Ethane has the lowest boiling point because it has the smallest molecules, with only two C atoms per molecule. This means its intermolecular forces of attraction will be weaker than in the other alkanes listed.

15

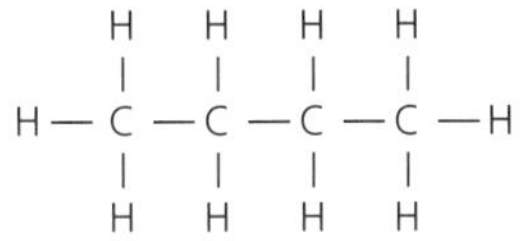

16 $C_3H_8 + Br_2 \rightarrow C_3H_7Br + HBr$

17 C_nH_{2n}

18

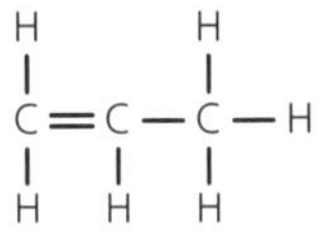

19 You would add the hydrocarbon to bromine water (or the other way round) and shake. If an alkene is present, the bromine water turns from orange to colourless.

20

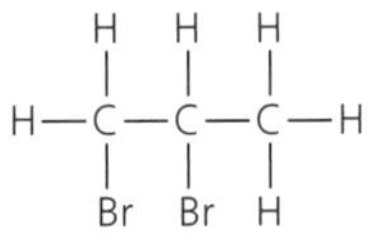

21 –OH, hydroxyl

22

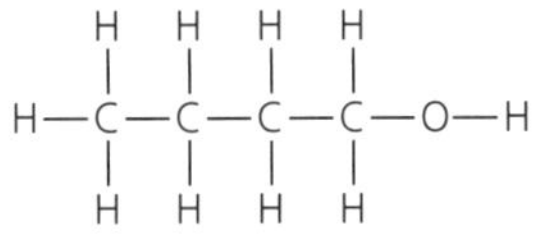

23 An optimum temperature of about 30 °C; no air can be present; the inclusion of yeast; the inclusion of sugar.

24 $C_2H_4 + H_2O \rightarrow C_2H_5OH$
This is an addition reaction because water is being added across the double bond.

25 Any answer between 0 and 6. The actual pH would depend on the concentration of the solution.

26 Methanoic acid

27

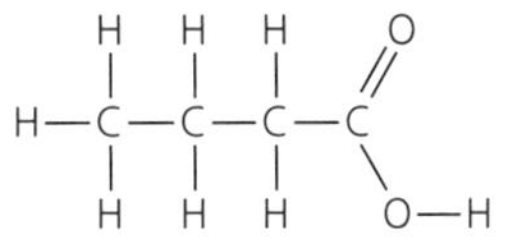

28 Lithium + ethanoic acid
→ lithium ethanoate + hydrogen

29

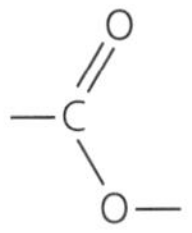

30 Methanol + ethanoic acid
⇌ methyl ethanoate + water

31 $CH_3CH_2OCOCH_2CH_3$ or $CH_3CH_2COOCH_2CH_3$

32

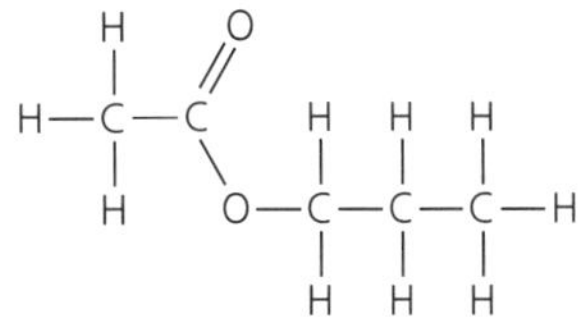

33 C=C

34 Poly(butene)

35 Phenylethene

36

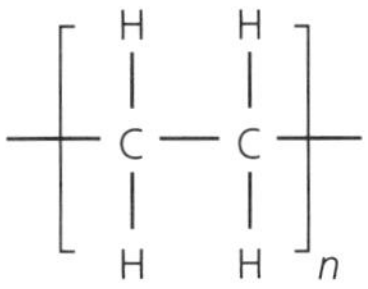

37 Addition polymers have no by-products when they are formed, but for every linkage made between two monomers in a condensation polymer, a small molecule (typically water) is produced as a by-product.

38 A dicarboxylic acid and a diol

Periodic Table

1	2												3	4	5	6	7	0
																		4 **He** Helium 2
7 **Li** Lithium 3	9 **Be** Beryllium 4												11 **B** Boron 5	12 **C** Carbon 6	14 **N** Nitrogen 7	16 **O** Oxygen 8	19 **F** Fluorine 9	20 **Ne** Neon 10
23 **Na** Sodium 11	24 **Mg** Magnesium 12												27 **Al** Aluminium 13	28 **Si** Silicon 14	31 **P** Phosphorous 15	32 **S** Sulfur 16	35.5 **Cl** Chlorine 17	40 **Ar** Argon 18
39 **K** Potassium 19	40 **Ca** Calcium 20	45 **Sc** Scandium 21	48 **Ti** Titanium 22	51 **V** Vanadium 23	52 **Cr** Chromium 24	55 **Mn** Manganese 25	56 **Fe** Iron 26	59 **Co** Cobalt 27	59 **Ni** Nickel 28	64 **Cu** Copper 29	65 **Zn** Zinc 30		70 **Ga** Gallium 31	73 **Ge** Germanium 32	75 **As** Arsenic 33	79 **Se** Selenium 34	80 **Br** Bromine 35	84 **Kr** Krypton 36
85 **Rb** Rubidium 37	88 **Sr** Strontium 38	89 **Y** Yttrium 39	91 **Zr** Zirconium 40	93 **Nb** Niobium 41	95 **Mo** Molybdenum 42	99 **Tc** Technetium 43	101 **Ru** Ruthenium 44	103 **Rh** Rhodium 45	106 **Pd** Palladium 46	108 **Ag** Silver 47	112 **Cd** Cadmium 48		115 **In** Indium 49	119 **Sn** Tin 50	122 **Sb** Antimony 51	128 **Te** Tellurium 52	127 **I** Iodine 53	131 **Xe** Xenon 54
133 **Cs** Caesium 55	137 **Ba** Barium 56	139 **La** Lanthanum 57 *	178 **Hf** Hafnium 72	181 **Ta** Tantalum 73	184 **W** Tungsten 74	186 **Re** Rhenium 75	190 **Os** Osmium 76	192 **Ir** Iridium 77	195 **Pt** Platinum 78	197 **Au** Gold 79	201 **Hg** Mercury 80		204 **Tl** Thallium 81	207 **Pb** Lead 82	209 **Bi** Bismuth 83	210 **Po** Polonium 84	210 **At** Astatine 85	222 **Rn** Radon 86
223 **Fr** Francium 87	226 **Ra** Radium 88	227 **Ac** Actinium 89 †																

H
Hydrogen
1

*58–71 Lanthanum series
†90–103 Actinium series

	a **X** b

a = relative atomic mass
X = atomic symbol
b = atomic number

	140 **Ce** Cerium 58	141 **Pr** Praseodymium 59	144 **Nd** Neodymium 60	**Pm** Promethium 61	150 **Sm** Samarium 62	152 **Eu** Europium 63	157 **Gd** Gadolinium 64	159 **Tb** Terbium 65	162 **Dy** Dysprosium 66	165 **Ho** Holmium 67	167 **Er** Erbium 68	169 **Tm** Thulium 69	173 **Yb** Ytterbium 70	175 **Lu** Lutetium 71
*	140 **Ce** Cerium 58	141 **Pr** Praseodymium 59	144 **Nd** Neodymium 60	147 **Pm** Promethium 61	150 **Sm** Samarium 62	152 **Eu** Europium 63	157 **Gd** Gadolinium 64	159 **Tb** Terbium 65	162 **Dy** Dysprosium 66	165 **Ho** Holmium 67	167 **Er** Erbium 68	169 **Tm** Thulium 69	173 **Yb** Ytterbium 70	175 **Lu** Lutetium 71
†	232 **Th** Thorium 90	231 **Pa** Protactinium 91	238 **U** Uranium 92	237 **Np** Neptunium 93	242 **Pu** Plutonium 94	243 **Am** Americium 95	247 **Cm** Curium 96	245 **Bk** Berkelium 97	251 **Cf** Californium 98	254 **Es** Einsteinium 99	253 **Fm** Fermium 100	256 **Md** Mendelevium 101	254 **No** Nobelium 102	257 **Lr** Lawrencium 103